Signal Processing in Noise Waveform Radar

For a complete listing of titles in the
Artech House Radar Series,
turn to the back of this book.

Signal Processing in Noise Waveform Radar

Krzysztof Kulpa

ARTECH HOUSE
BOSTON | LONDON
artechhouse.com

Library of Congress Cataloging-in-Publication Data
A catalog record for this book is available from the U.S. Library of Congress.

British Library Cataloguing in Publication Data
A catalogue record for this book is available from the British Library.

Cover design by Vicki Kane

ISBN 13: 978-1-60807-661-1

685 Canton Street
Norwood, MA 02062

10 9 8 7 6 5 4 3 2 1

To my beloved wife Ewa

Contents

Acknowledgments

I thank all my colleagues, coworkers, and students whose work has contributed to the field of noise radar and this book. Among the many I thank are Dr. Mateusz Malanowski, Dr. Jacek Misiurewicz, Dr. Piotr Samczynski, Lukasz Maslikowski, and my university team for their hard work on algorithm development, creative involvement in measurement campaigns, and long hours on processing simulated and recorded data. I also thank Professor Konstantin Lukin and Professor Bronislaw Stec for their interesting discussions, and for sharing with me the images and measurement results presented in this book.

1

Introduction

The concept of radar—radio detecting and ranging—emerged in the beginning of the twentieth century. The father of radar was Christian Huelsmeyer, who applied for a patent for his "telemobiloscope" on April 30, 1904. His device worked quite well and detected ships at ranges up to 3 km, but he had no success in selling telemobiloscopes, which is why the early radar concept faded from memory. The reinvention of radar was done almost simultaneously in many countries in the 1920s and 1930s, and great progress in radar technology was made during World War II. In the last 60 years, many different radar technologies have been developed, increasing the radar's ability to detect targets in the presence of ground and sea clutter, to measure the target range velocity, and to create high-resolution images. In military areas radars are used for detecting, tracking, and imaging enemy airplanes, ships, satellites, and vehicles. They are used also for fire control, missile guidance, proximity detections, target identification, and many other purposes. Radars are also found in many civilian applications. They are used in air traffic control (ATC), ground traffic monitoring (including enforcement of the speed limit by police), collision avoidance and traffic assistance (automotive radars), observation of the Earth (biomass calculation, disaster control, change detection, and ocean currents monitoring) and in industry (measurement of liquid or powder level in tanks, object positioning, movement detection, and so on).

Early radars were noncoherent pulse radars. To obtain a long-range detecting capability, the radar emitted very short electromagnetic pulses with very high peak power (up to several megawatts), and such a large peak caused a lot of difficulties. The transmitter was big and heavy. Microwave waveguides had to conduct megawatt pulses, and their construction was complicated. The large peak power also had an impact on the environment. All life in the surrounding areas near the radar was exposed to this hazard; birds were often killed by electromagnetic radiation, and trees sustained damage.

The maximum detection range of a radar is limited by several factors such as the antenna, the size of the target, the noise level in the receiver, and, one of the most important, the total energy radiated towards the target. It is possible to exchange the transmitted pulse width with the peak power. In a classical pulse radar an increase of the pulse length decreases the radar range resolution, which is generally undesirable. To obtain relatively low peak power and a high-range resolution simultaneously, different types of pulse compression are used. Among them the most popular has been compression based on linear frequency modulation. To improve the radar's properties, different types of intrapulse modulation have been investigated, such as binary-phase modulation, polyphase modulation, and noise or pseudo-noise modulations.

In military radars there is a continuous contest between the radars and electronic support measurement (ESM) devices. The modern military radar should detect the target before being detected by an enemy ESM receiver. The ESM detection range depends on the peak power of the signal transmitted by the radar, so the concept of low probability of interception (LPI) continuous-wave radars was introduced. Such a radar, usually equipped with separate transmitting and receiving antennas, could simultaneously transmit and receive signals so the pulse length can be extended almost infinitively (ranging in practice from nanoseconds to seconds, several orders of magnitude). As a result, the peak transmitted power could be reduced by even a million times (60 dB). Again, as in the case of pulse compression, different types of signal modulations have been investigated. Many continuous-wave radars used sawtooth frequency modulation of the transmitted signal. The signal processing algorithms for linear frequency-modulated continuous-wave (LMCW) radars are relatively simple (based on the two-dimensional FFT) and require reasonably low computational power, so LMCW radars could be simple and cheap.

Recent progress in ESM technology has made it possible to detect FMCW radars, so radar scientists are looking for new technologies that will improve the LPI properties of the radar and prevent ESM receivers from detecting and classifying the signal transmitted by the radar. As of the time of this writing, there were several candidates for modern LPI radars. The first of them was noise radar technology. The second was passive coherent location technology, and the third was passive technology based on a signal's time difference of arrival (TDOA) at several different locations.

In most classical active radars the transmit signal is periodic. The periodicity introduces ambiguities in the radar measurements. The long-range pulse radar sends a sounding pulse and waits for an echo to return from the maximal distance. As a result, the pulse repetition frequency is low and such a radar suffers from Doppler ambiguity (Doppler frequency measurement is ambiguous).

It is also possible to build high pulse repetition frequency (high PRF) radars, but an increase of the PRF decreases the unambiguous detection range and, as a result, introduces range ambiguities. It was not possible to measure both the distance from the target and the target radial velocity instantaneously without the ambiguity caused by the sampling theorem. To fulfill different requirements, three main types of pulse radar are used: low pulse repetition frequency (LPRF), medium pulse repetition frequency (MPRF), and high pulse repetition frequency (HPRF). To resolve Doppler or range ambiguity, different estimation methods, usually in conjunction with pulse staggering (changes of PRF during target illumination), are used. LPRF radars are free from range ambiguity, but have Doppler ambiguity. They are commonly used as ground-based surveillance radars. HPRF radars have range ambiguity but do not have Doppler ambiguities. They are commonly used as airborne radars and tracking radars. MPRF radars have both range and Doppler ambiguities. This radar operation mode is often used as an additional mode in HPRF airborne radars. Nevertheless, the idea of the pulse radar can lead to a relatively simple design, which can be implemented using analog-only devices.

The rapid progress in digital signal processing (DSP) hardware and algorithms has enabled designers to use more sophisticated ideas and more complicated signal processing algorithms in modern radars. One such idea is to use a continuous wave (CW) instead of high power pulses. This idea is not new; the first Daventry experiment (see Chapter 2 for details) was based on CW radio emissions, but the practical use of CW emissions was limited by the necessity to build a sophisticated matched filter set. Modulated CW radars were built during World War II and discussed in the earliest postwar literature, but most early CW radars were constructed as Doppler-only radars (e.g., police radars) to measure the velocity of the target. More advanced systems used periodical linear frequency modulation (FMCW) to obtain range resolution. Due to a lack of modulation (in single-frequency police radars) or periodicity in modulation (in FMCW radars), the ambiguity problem was not removed, but CW radars have some very strong advantages. Many CW radars use low transmitted power (1 mW up to 100W) and due to low peak power (equal to the mean transmitted power), it is difficult to detect and classify this type of radar, so they are treated as a low probability of intercept (LPI) class of radars. Such radars can detect targets before being detected.

The idea of LPI radars in connection with the requirement of designing an ambiguity-free radar leads directly to the concept of the noise or pseudonoise radar. In this kind of radar the target is illuminated by continuous noise-like radiation. The reflected power is collected by the radar receiving antenna, and detection is based on the matched filtering of the received signals. Due to the fact that both the target range and the target velocity are unknown, then instead of using a single filter matched to the selected target's range and velocity, it is necessary to use a two-dimensional set of filters, matched to all possible targets' range-velocity pairs. The detection of targets in noise radars thus requires very high computational power and cannot be effectively implemented in the analog technique.

The theoretical cross-ambiguity function of the noise signal has a delta shape: infinite value at zero time shift, zero Doppler frequency, and zero elsewhere. The transmit signal is not periodic, so there are no range or Doppler ambiguities in the measurement.

Of course, nothing is free, so we pay a price for the lack of ambiguity. The delta-shaped cross-ambiguity function is for an ideal white noise signal, infinite in time and frequency (bandwidth). Such a signal would have infinite energy and could not be generated by any real radar. In practice we can generate a sounding system limited in time and bandwidth (finite energy). In such a case the cross-ambiguity signal has a clearly visible peak at the zero Doppler frequency and zero time delay, and the width of the peak in time is equal to the inverse of the signal bandwidth and in frequency, the inverse of signal duration time. Such a signal also has nonzero values out of the peak region. Time and range residues of the cross-ambiguity function (which can be treated as the sidelobes in the case where single noise realization is treated as a deterministic signal, for example, stored in memory) have random nature, but the level is BT (the product of the signal bandwidth and duration time) below the peak value. This means that the ambiguity problem was converted to the masking problem. Sidelobes of strong echoes can mask weak ones if the ratio of the strong to weak echoes is higher than the time bandwidth product.

The limitation of weak-signal detection in a noise radar is due to the problem of the strong echo signals, which usually originate from nearby ground clutter and are received simultaneously with the weak target echo. To ensure correct detection of distant targets, the radar has to have a very high dynamic range and very low intermodulation. When the matched filter of the noise radar is tuned to the range and Doppler velocity of the weak echo, the power of all other echoes can be summed with the receiver thermal noise power decreasing the effective signal-to-noise ratio of the weak target. This phenomenon, together with the method of countering it, is described in Chapter 4.

The concept of noise radar may be used for the detection of moving targets, but it can also have many other applications. There are several papers showing the possible implementation of noise technology in an imaging radar working in the synthetic aperture mode or inverse synthetic aperture mode (ISAR). This concept can be extended to the passive detection, imaging, and identification of targets. Noise radar will probably be used in the future in other fields, such as air traffic control, pollution control, and security applications.

This book presents the basic idea of noise radar and details of signal processing. It is assumed that the reader has a basic knowledge of signal processing

[1–3] and general radar technology [4–9]. However, in Chapter 2 the fundamentals of the classical pulse radar are presented. More detailed analyses of classical radar processing can be found in the radar engineering bible [4] and other radar books [5–9]. In Chapter 3 the basic noise radar concept and correlation processing are introduced. Chapter 4 presents the effect of masking weak echoes by a strong cross-talk signal and strong echoes. In this chapter the number of algorithms designed to mitigate this problem is discussed. In Chapter 5 the idea of the multistatic noise radar network is presented. This chapter presents the idea of placing the transmitter and receiver in different locations, and discusses the advantages and limitations of such an approach as well as the methods for synchronizing the transmitter and receiver. This chapter also describes a more advanced idea of the use of a multistatic system, where several transmitters illuminate the target using the same carrier frequency. The approach presented leads to an increase of surveillance volume, measurement accuracy, and probability of detection.

In Chapter 6 the principles of noise synthetic aperture radar are presented. In Chapter 7 the new idea of the passive detection and imaging of moving targets is described. The idea presented is based on radiometry and exploits the thermal emission of microwave noise signals originating from moving targets. Chapter 8 shows selected examples of noise radars and draws conclusions on noise radar technology.

References

[1] Oppenheim, A. V., and R. W. Schafer, *Digital Signal Processing,* Upper Saddle River, NJ: Prentice Hall, 1975.

[2] Lyons, R. G., *Understanding Digital Signal Processing,* 3rd ed., Upper Saddle River, NJ: Prentice Hall, 2010.

[3] Smith, S. W., *The Scientist & Engineer's Guide to Digital Signal Processing,* 1st ed., San Diego, CA: California Technical Publications, 1997.

[4] Skolnik, M. I., *Radar Handbook,* 3rd ed., New York: McGraw-Hill, 2008.

[5] Skolnik, M., *Introduction to Radar Systems,* 3rd ed., New York: McGraw-Hill, 2002.

[6] Wehner, D. R., *High Resolution Radar,* 2nd ed., Norwood, MA: Artech House, 1995.

[7] Curlander, J. C., and R. N. McDonough, *Synthetic Aperture Radar Systems and Signal Processing,* New York: John Wiley & Sons, 1991.

[8] Franceschetti, G., and R. Lanari, *Synthetic Aperture Radar Processing,* Boca Raton, FL: CRC Press, 1999.

[9] Richards, M. A., J. A. Scheer, and W. A. Holm, *Principles of Modern Radar: Basic Principles,* Raleigh, NC: SciTech Publishing, 2010.

2

Radar Principles

A radar is a device that can remotely detect targets and measure the distance. The name radar comes from acronym of the technology called *radio detection and ranging*. Classical surveillance radar scans the area using a single antenna beam or a fan of beams in which short (and very often coded) radio-frequency pulses are emitted. The signal reflected from the target is received by the radar antenna and is then further processed by the receiver and radar signal processing units to extract information about the target and its location. The first radio-detection devices were developed by German engineer Christian Hülsmeyer. He started his work after witnessing a collision of two ships. His device, which he named the telemobiloscope, was the predecessor of the modern radar. The drawings of the telemobiloscope, as presented in his German patent [1] are presented next. The general idea is presented in Figure 2.1. Ship A is equipped with a radar on the mast, illuminating ship B, and the reflected signal is received by a radar receiver.

Technical details of the invention are presented in Figure 2.2. It consisted of a Righi-type spark gap radio-wave transmitter (as a discharging device), operating at a wavelength of 40 to 50 cm (600–750 MHz). The radio-wave pulses were directed to the target using a funnel-shaped reflector and a narrowbeam, horn-type antenna. The second narrowbeam receiving antenna concentrated the energy reflected from the target onto the detector, which was based on a coherer. The two antennas were separated by a metal shield and mechanically scanned in a horizontal direction. As an indicator, an electric bell was used, and it rang when the antennas were pointed towards the target.

The mechanical scanning device was used to find the direction of the target. The idea of this mechanism is shown in Figure 2.3.

In June 1904 Hülsmeyer presented the new anticollision devices to the director of a Dutch shipping company, showing that it was possible to detect ships

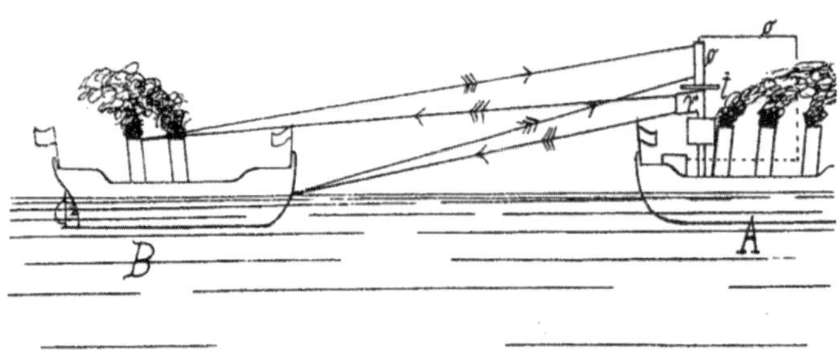

(b) (a)

Figure 2.1 The idea of the first radar telemobiloscope as presented in Christian Hülsmeyer's patent. Ship A is equipped with a radar on its mast, illuminating ship B, and the reflected signal is received by a radar receiver. (From [1].)

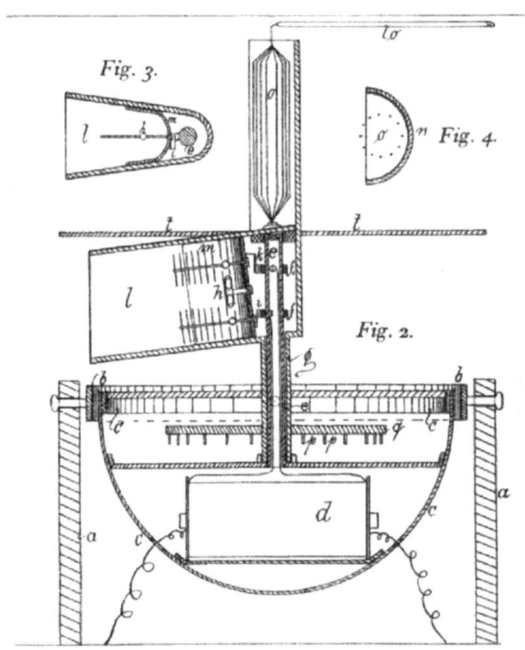

Figure 2.2 Technical details of the invention. (From [1].)

at a distance of up to 3 km. The inventor also planned to add ranging capability to his devices, using two-dimensional scanning. He obtained several patents on the telemobiloscope in Germany, Great Britain, and the United States. The U.S. Patent No. 810,150 was dated January 16, 1906. Hülsmeyer founded the Telemobiloscop-Gesellschaft Hülsmeyer und Mannheim company, but his invention was too far ahead of its time. At that time people believed more in

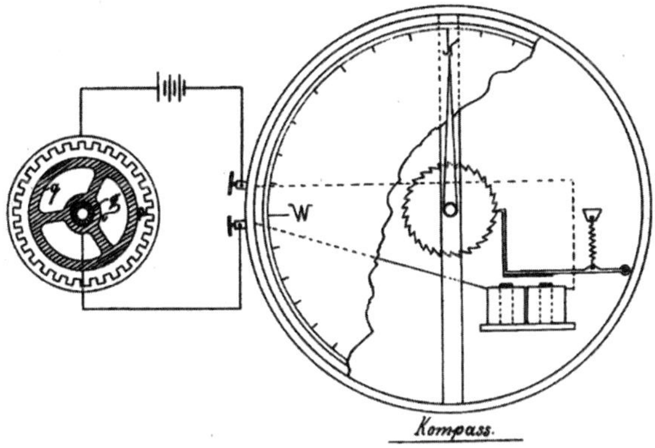

Figure 2.3 Scanning mechanism. (From [1].)

optical observation than in technical support, and bad weather conditions were not taken into consideration so seriously. However, his device was able to work 24 hours a day, making observation in day and night conditions, and also was able to observe the target on foggy days. His invention could have saved the *Titanic* and many souls. As a result, Telemobiloscop-Gesellschaft Hülsmeyer und Mannheim had no commercial success and the radar, or more specifically the telemobiloscope, concept was forgotten for more than 20 years.

In fact, the telemobiloscope was the first noise-pulse radar, while the sounding pulses, produced by an electrical discharge filtered by a bandpass antenna system, were band-limited noise pulses from the present point of view.

Progress in the construction of radio devices, power amplifiers, receivers, and selected filters was the reason why many scientists in the early 1920s started to carry out independent research on radar technology. In 1922 Guglielmo Marconi presented his idea of detecting remote objects by radio signals, but it took him 11 years to build his radar demonstrator. The first radio range-finding experiment took place in Great Britain in 1924, when Sir Edward Victor Appleton used radio echoes to measure the height of the ionosphere. The next important contribution to radar technology was made by the British Post Office issuing a report in 1932, which pointed out that airplanes were interfering with radio signals and reradiating them.

In 1935 Sir Robert A. Watson-Watt worked on the concept of a "death-ray" weapon. During his research he discovered that the death-ray concept, very popular in science fiction literature at that time, was not yet possible, because the energy required to kill the pilot of the airplane was far beyond the technical limitations. So instead of discovering a directed-energy weapon, he found that by using a radio transmission, it was possible to detect an echo from a plane at a range exceeding 200 km. On February 26, 1935, he carried out an experiment at Weedon, 6 km southeast of Daventry (a town located between London and

Birmingham) and obtained a detection of an RAF Heyford bomber at a height of 1,800m and a distance of 8 miles, using a radio beam from the short wave BBC transmitter at Daventry, working at the 6-MHz frequency band (49m wavelength). It was the first passive radar experiment in which a noncooperative transmitter of opportunity was used. The problem of direct signal cancellation was solved by using two receiving antennas placed at positions that allowed the direct signal from the two antennas to be summed in an opposite phase and thus strongly attenuated (see Figure 2.4). The echo signal coming from other directions was amplified and displayed using a cathode ray tube (CRT) oscilloscope. On the basis of that experiment, the concept of active radar was developed in Great Britain. This concept was shortly converted to a chain of active pulse radars, which became operational in 1937 under the name of Chain Home. Before World War II, 10 countries (Britain, France, Germany, Hungary, Italy, Japan, the Netherlands, Switzerland, Russia, and the United States) had operational radars in their inventories.

2.1 Pulse Radar and Its Range Equation

In the Daventry experiment continuous-wave illumination was used. The continuous-wave illuminator of opportunity was chosen, but at that time it was impossible to estimate the distance to the target using such an illuminator. Even now, using modern digital signal processing techniques, it would be very difficult to estimate the range to the target using such illumination, and effective range resolution would be of dozens of kilometers. The scientists of the 1930s developed an alternative idea. Instead of providing continuous illumination, they used short pulses to illuminate the target and they constructed pulse radar. This technology is also used in many modern radars.

The idea of pulse radar is very simple. Pulse radar emits a train of short, high-power electromagnetic pulses, and after each emission starts to listen to the pulse echoes. The distance to the object is obtained by measuring the time delay between the emitted pulse and the return echo, whereas the target position is estimated by finding the direction from which the returned echoes are the strongest. Let us consider the mathematical details of the radar principles.

The pulse radar emits high-power, short electromagnetic pulses at a given carrier frequency F, using a directional antenna of gain G_T. The antenna directive gain is obtained by emitting all electromagnetic energy within a narrow angle instead of emitting it omnidirectionally. The antenna gain in a simple case (with an ideal, lossless antenna) can be calculated as ratio $G_T = 4\pi/\alpha$, where α is the solid angle of illumination.

The pulses are reflected by a target (see Figure 2.5) and are received by a receiving antenna. In most cases the same antenna is used both for transmitting

Figure 2.4 Sketch of the Daventry experiment, February 26, 1935, set up by A. F. Wilkins to detect radio signals reflected from an aircraft.

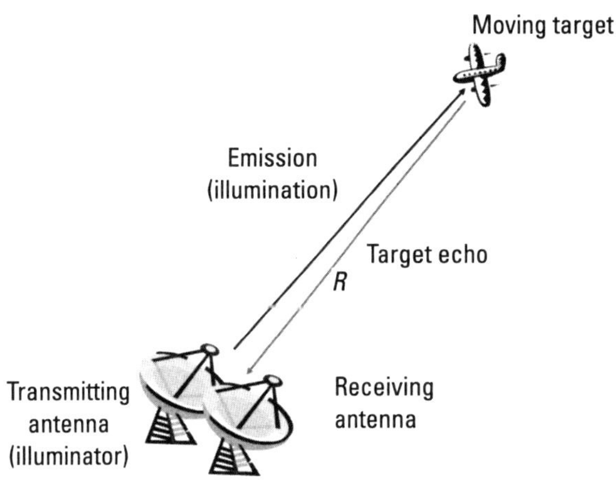

Figure 2.5 Monostatic radar scenario with separate transmit and received antennas.

and receiving, allowing the radar to be equipped only with one antenna, as presented in Figure 2.6. The transmit signal is separated from the received one using a microwave circulator and/or transmit/receive switch now constructed using PIN diodes.

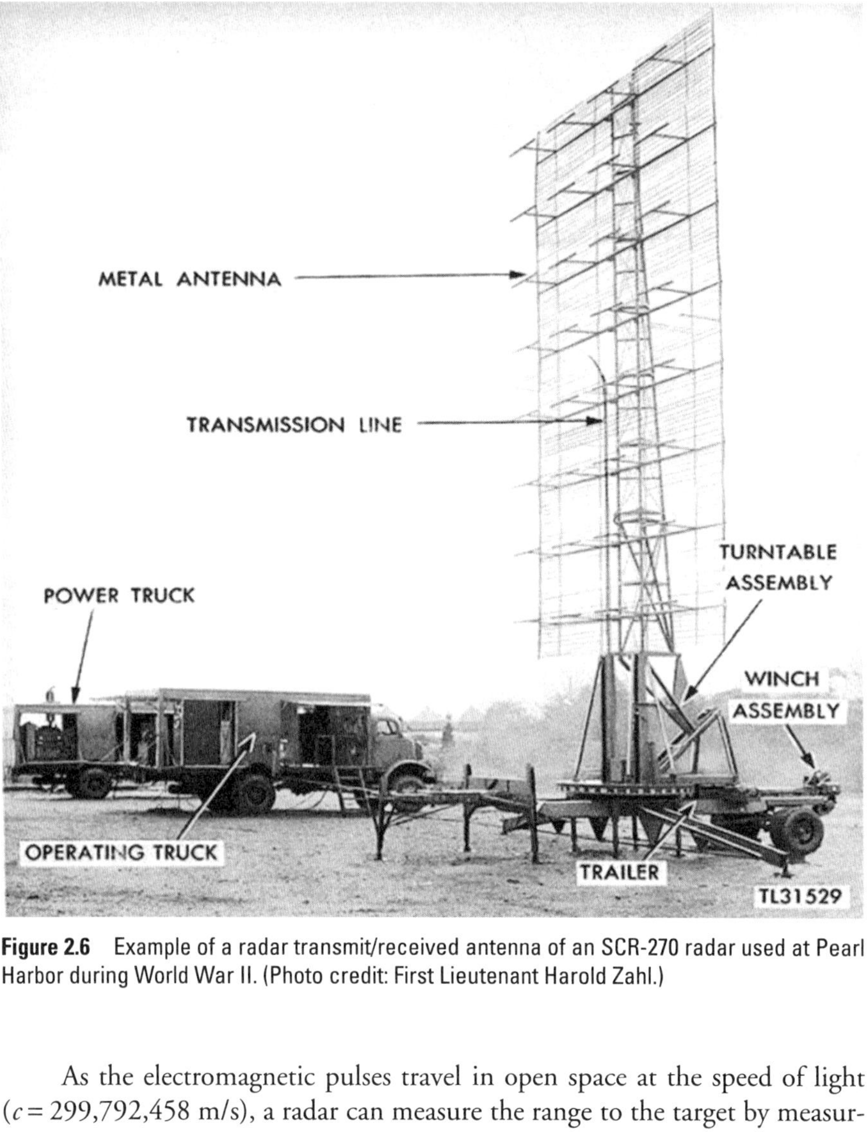

Figure 2.6 Example of a radar transmit/received antenna of an SCR-270 radar used at Pearl Harbor during World War II. (Photo credit: First Lieutenant Harold Zahl.)

As the electromagnetic pulses travel in open space at the speed of light ($c = 299{,}792{,}458$ m/s), a radar can measure the range to the target by measuring the time delay between the emitted and received pulses, and target angular coordinates, determined in selected coordinate system, are estimated by finding the maximum of the echo power received during the scanning process (see Figure 2.7).

Now let us consider the power relations assuming that the radar emits short pulses of power P_I towards the target. If an omnidirectional antenna were used, the power density at the range R would be the ratio of emitted power to the surface of the sphere of radius R. The use of the antenna of directional gain $G_T = 4\pi/\alpha$ will increase the power density at a target G_T times and the final power density will be described by the following formula [2]

$$p(R) = \frac{P_T G_T}{4\pi R^2} \tag{2.1}$$

The total power illuminating a target is the product of the power density and effective cross-section of the target S_o and can be described as

$$P_S = \frac{P_T G_T}{4\pi R^2} S_o \tag{2.2}$$

The energy that illuminates the target can be reflected, absorbed, or passed through a target.

In our case we assume that all energy is reflected omnidirectionally by the target, which is true only for a small spherical target. In the case of a more complex target, we can modify S_o to incorporate the angle relation of scattering properties, absorption, and other effects. Then, keeping in mind the properties of S_o, we can calculate the power received by the radar. The unit of radar cross-section is 1 square meter. Although the range equation has multiplicative terms, electronic engineers prefer to use decibel units instead of linear ones, because in such cases equations take the additive form. So instead of using an [m^2] unit for radar cross-section, decibel units are often used with respect to 1 square meter, and such a unit is called dBsm (decibels related to one square meter). Table 2.1 presents the relation between those units and example targets of similar size. It is important to remember that the target radar cross-section depends significantly on the radar wavelength, aspect ratio, and material from which the target is constructed.

Let us assume again the ideal case, in which the energy collected by the target is reradiated omnidirectionally. The reradiated energy illuminates the whole sphere of radius R centered on the target, and the radar receiving antenna

Table 2.1

Radar Cross-Section in Square Meters and dBsm

m^2	dBsm	Example
0.001	–30	Stealth airplane
0.01	–20	Small UAV, bird
0.1	–10	Rocket
1	0	Fighter, standing man
10	10	Small boat
100	20	Passenger plane
1,000	30	Ship

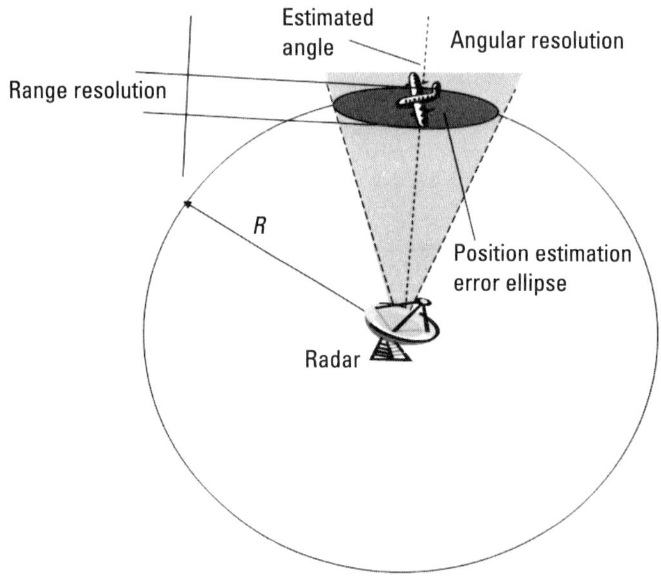

Figure 2.7 Target position estimation and radar resolution.

of an effective surface S_R collects the energy proportional to the ratio $S_R/4\pi R^2$. The received energy can be described by the following formula

$$P_R = \frac{P_T G_T}{16\pi^2 R^4 L} S_o S_R \tag{2.3}$$

where L denotes all insertion losses in the radar system, including transmission losses, propagation losses and receiving losses. By substituting the antenna effective surface with the receiving antenna gain G_R in (2.3) one can obtain the classical radar equation

$$P_R = \frac{P_T G_T G_R \lambda^2}{(4\pi)^3 R^4 L} S_o \tag{2.4}$$

where $\lambda = c/F$ is the wavelength of the transmitted signal.

To detect the target, it is necessary to measure the received power and to make the decision that the received power originates from the target of interest. Usually, the received power is very low and an analog amplifier has to be used.

In free-space conditions the received echo signal is the only signal received by the antenna, but the analog amplifier produces the noise signal. The goal is to distinguish between the amplifier noise and received echo by choosing between two hypotheses: H_0 (there is only noise in the signal) and H_1 (there is the echo plus noise).

The radar echo detection is equivalent to the classical problem of detecting a known signal buried within white noise [3]. The receiver's equivalent noise power can be expressed as

$$P_N = kT_R B \tag{2.5}$$

where T_R denotes the effective system noise temperature (dependent on the temperature of the receiver, the receiver's noise figure, antenna noise, and outer space noise), B denotes the receiver bandwidth (assuming the matched filtering of a received signal), and k is Boltzmann's constant ($1.3806505 \times 10^{-23}$ [J K^{-1}]). Using the Neyman-Pearson criterion, it is assumed that hypothesis H_1 is true (there is an echo in the signal), when the echo power is higher than the noise power multiplied by the detectability factor D_o, usually having the value of 12–16 dB, depending on the assumed probability of a false alarm. The radar detection criterion can then be written as

$$\frac{P_T G_T G_R \lambda^2}{(4\pi)^3 R^4 L} S_o > kT_R B D_o \tag{2.6}$$

The maximum detection range is the most important parameter of a radar. It can be predicted by the formula

$$R_{\max} = \sqrt[4]{\frac{P_T G_T G_R \lambda^2 S_o}{(4\pi)^3 L k T_R B D_o}} \tag{2.7}$$

The formula (2.7) is the simplified version of the radar range equation, where all losses (including propagation losses) are aggregated in one system loss labeled L. For pulse radar using a matched filter to eliminate out-of band interferences and noises, the receiver bandwidth B is inversely proportional to the pulse width t_p. Replacing the receiver bandwidth with pulse duration t_p in (2.7), one can obtain

$$R_{\max} = \sqrt[4]{\frac{E_T G_T G_R \lambda^2 S_o}{(4\pi)^3 L k T_R D_o}} \tag{2.8}$$

In high-frequency, long-range radars, the insertion losses depend on the radar-target range, and the maximum detection range must be calculated numerically. It depends strongly on weather conditions, especially air humidity. However, in this book this effect is not taken into consideration, and for the sake of simplicity, (2.8) is used.

As stated in (2.8), the detection range depends only on transmitting pulse energy, transmitter and receiver antenna gains G_T, G_R, target cross-section S_o, and wavelength λ, and does not depend on the pulse length or the receiver bandwidth (assuming that a matched filter is used). Formula (2.8) describes a single-direction/single-pulse case. It is easy to see that this equation is also a good approximation of the detection range for more complex detection schemas, using the multipulse technique with coherent integration or continuous waveforms. In that case E_T stands for total energy emitted towards a target, assuming that fully matched filtering is possible (the signal can be integrated coherently). It is not always true, however, because many radars use noncoherent (amplitude-based) signal integration. As well, in such cases, (2.8) can be very useful because all losses such as integration losses, filter mismatched losses, antenna losses, and others can be calculated and incorporated in loss factor L.

The two main types of radars are surveillance radar and tracking radar. Surveillance radar searches the surveillance space volume and detects all targets. The radar detections (plots) are used for updating existing the track position (track-while-scan function) and to initiate a new track. Tracking radar has another goal: to provide the position information of a single selected target so that all available energy is allocated to detect and track that target. The target illumination time can be much longer than in surveillance radar and is limited by the required position data refresh rate.

Modern radars are multifunctional, often equipped with electronic scanning antennas (see Figure 2.8) and such a radar can allocate part of its resources (time and energy) for surveillance scanning and the rest for tracking multiple selected targets. For all of these processes, (2.8) can also be used to predict the detection and tracking distance.

As mentioned before, a wide variety of radars work in the surveillance mode. Let us consider a surveillance radar equipped with a transmitter having the mean transmitted power P_{Tm}, searching the surveillance solid angle α_s in time t_s. In many cases the surveillance solid angle is a whole semisphere (2π).

The idealized transmitting antenna with gain G_T emits electromagnetic waves in the solid angle $\alpha_T = 4\pi / G_T$. In order to scan the whole surveillance volume (surveillance angle), it is necessary to scan ns $= \alpha_s / \alpha_T = \alpha_s G_T / 4\pi$ directions, so the time of illuminating a single direction (called time on target) is limited to the value

$$t_t = \frac{t_s}{n_s} = \frac{4\pi t_s}{\alpha_s G_T} \tag{2.9}$$

Figure 2.8 Electronic scanning antenna of the Cavalier Air Force Station Perimeter Acquisition Radar Characterization System.

The total energy associated with a single scan direction is equal to $E_T = t_t P_{Ts} = \dfrac{4\pi t_s P_{Ts}}{\alpha_s G_T}$, where P_{Ts} is the portion of the multifunction radar's average power assigned to the surveillance function. Consequently, the surveillance detection range can then be expressed as

$$R_{\max} = \sqrt[4]{\frac{P_{Ts} t_s G_R \lambda^2 S_o}{(4\pi)^2 \alpha_s L k T_R D_o}} = \sqrt[4]{\frac{P_{Ts} t_s S_R S_o}{4\pi \alpha_s L k T_R D_o}} \tag{2.10}$$

As follows from the analysis of formulas (2.9) and (2.10), the detection range depends on the receiving antenna gain, mean transmitter power, scanning angle, and time. The gain of the transmitting antenna does not contribute to the detection range, but it influences the time on target. Time on target for surveillance radar is described by (2.9), but other factors can also influence it. The other limiting factors are time of the electromagnetic wave propagation (time in which electromagnetic pulses propagate from the radar to the target and back) and the time the object stays in a radar range-angle resolution cell.

In many radar systems the same antenna is used for transmitting and receiving. It means that $G_T = G_R$ and all transmitted energy is collected by the receiving antenna having the same beamwidth. To extend the effective radar range, it is possible to use a separate receiving antenna with higher gain than the transmitting antenna, but the use of a single receive antenna will provide mismatched antenna pattern losses when the receive antenna beamwidth is smaller than the transmit one and part of the transmit energy is lost. To avoid such losses, a set of more directional, high-gain antennas can be used. This leads directly to the concept of a multibeam or stacked-beam antenna presented in Figure 2.9, which is very popular in 3-D surveillance radar. This idea can be further extended, and it is possible to develop a system with an omnidirectional transmit antenna and a multibeam receive antenna set.

2.2 Radar Range Measurement and Range Resolution

The range R from a radar to the target is determined by measuring the time in which the electromagnetic pulse is propagated from the radar to the target and back. The time delay between the transmitted and received signal τ is equal to

$$\tau = \frac{2R}{c} \tag{2.11}$$

where c is the velocity of light, equal to 299,792,458 [m/s] in a vacuum. In practical cases the radar pulses travel in the Earth's atmosphere, and the velocity of light depends slightly on the air temperature, pressure, and humidity, but such aspects are not considered in this book and standard light velocity will be used. Thus, the range to the target R can be easily calculated when the time delay is known, using the following formula:

$$R = \frac{\tau c}{2} \tag{2.12}$$

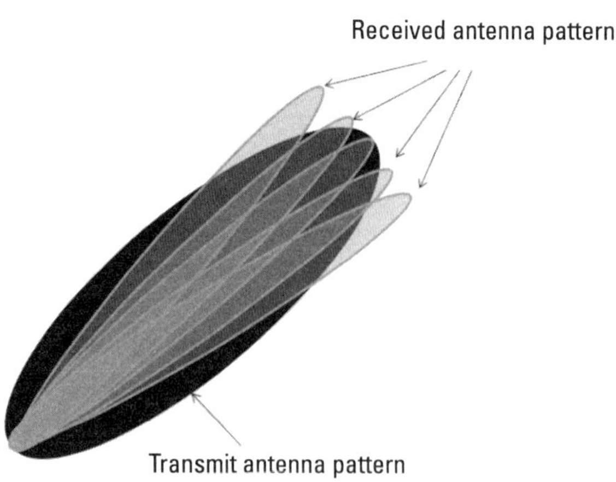

Figure 2.9 Multibeam antenna concept.

In other words, the problem of range estimation can be converted to the problem of estimation of time delay . Of course, the time delay can be estimated by means of different methods. The most popular is finding the coordinates of the maximum value of the cross-correlation function between the transmitted signal x_T and received signal x_R, defined as

$$r(\tau) = \int x_R(t) x_T^*(t-\tau)dt \tag{2.13}$$

where * denotes a complex conjugation. Usually the formula (2.13) is applied to the downconverted signals shifted to the baseband.

In a classical pulse radar, electromagnetic energy is emitted as a short pulse of a sine wave signal of frequency F and duration time t_p (see Figure 2.10). To maximize the received signal-to-noise ratio, matched filters are often used in radar technology. The matched filter generates at the output a signal proportional to a cross-correlation function determined between the transmitted and received signals. In this case the output signal has a triangular envelope of total time duration $2t_p$, filled with a carrier frequency cosine signal (see Figure 2.10). The basic question related to the range measurement is, what are the radar range resolution and accuracy of range measurements? Resolution is the ability to resolve two closely located targets and the question is, how far from

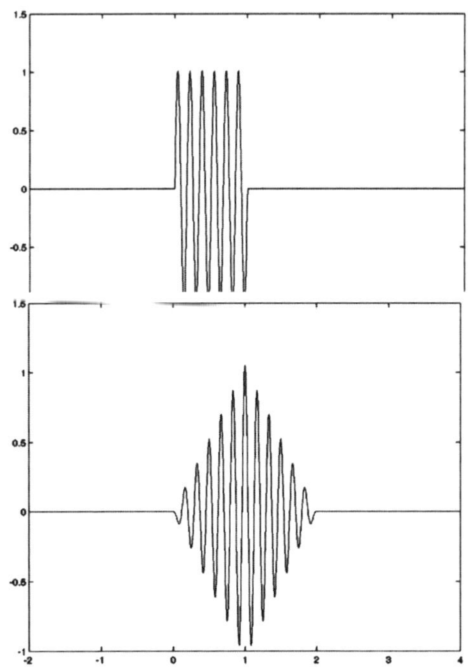

Figure 2.10 The received pulse after matched filtering.

each other should two point targets be (usually defined as targets of equal radar cross-section) in order to be resolved by radar (detected as two targets, not one)? The second question is, how possible is it to precisely measure the distance to a single point target using a selected waveform?

While the filter matched to the pulse t_p has a bandwidth of approximately $B = 1/t_p$, it is easy to calculate the range resolution for different pulse durations. Example calculations are presented in Table 2.2.

The echoes received from the two separate objects are fully separated if their time delay difference is equal to or higher than $2t_p$, which means that their radial distance is higher than t_pc. In cases when the time delay is smaller, the two echo components will overlap at the output of the matched filter. The shape of the output signal will depend on the difference of the phase of the echo signals.

In Figure 2.11 the shapes of the signal at the output of the matched filter are presented for a phase difference of $\varphi = 0°$, 90°, 180° and the time delay between the signals t_p, $1.5t_p$, and $2t_p$, respectively. It is usually assumed that two objects can be distinguished if their range separation is greater than $t_pc/2$, but full separation is obtained for ranges greater than t_pc.

The spectrum of a rectangular pulse signal is described by a *sinc* function, presented in Figure 2.12. The main lobe spectrum width (at the -3-dB level) is almost equal to $B = 1/t_T$. Naturally, the total signal spectrum is spread over a much wider bandwidth, and the first sidelobes are attenuated only by 13 dB. To avoid interferences between radars working at different frequencies, the transmitted spectrum is sometimes limited by using frequency filtering or time windowing. An example of limiting the transmitted signal spectrum by applying a Hamming window to the transmitting signal is given in Figure 2.12. This approach is seldom used in practice, as it requires that the transmitter operate in the linear mode, greatly reducing the peak transmitter power relative to saturated operation. Thus, to limit the frequency sidelobes, microwave filters are often used. Such filters reduce frequency sidelobes but also introduce additional losses (0.5–3 dB), and widen the main lobe (10%–40%), therefore decreasing the radar range resolution and accuracy. In that case the correlation function

Table 2.2
Radar Pulse Width, Bandwidth, and Range Resolution

Pulse t_p [ns]	Bandwidth $B = 1/t_p$ [MHz]	Range Resolution ΔR [m]	Example
1	1,000	0.15	Imaging radars, SAR/ISAR
10	100	1.5	Combat radars, battlefield radars
100	10	15	Maritime patrol radars, military radars
1,000	1	150	ATC radars (airport radars)

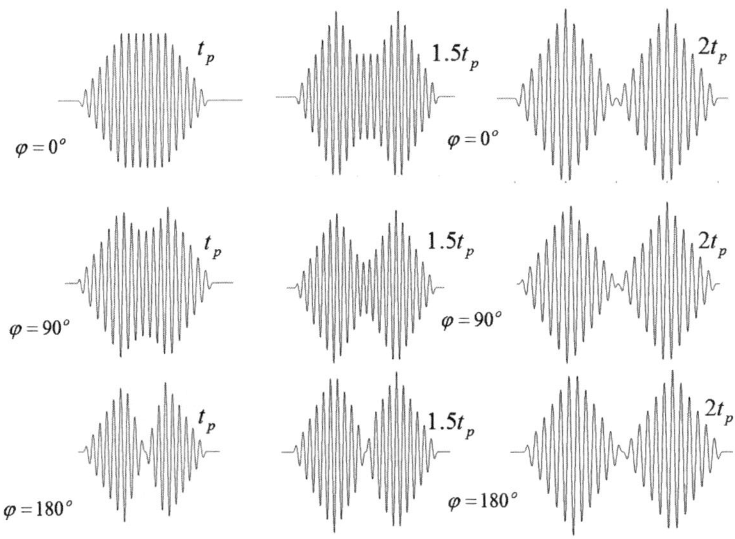

Figure 2.11 The two echoes after matched filtering.

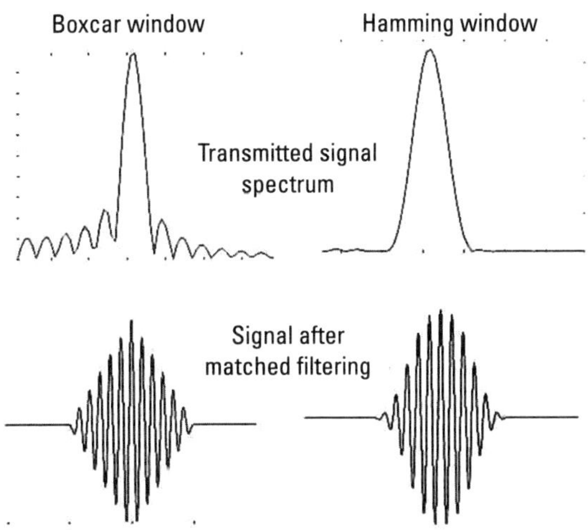

Figure 2.12 The transmitted signal spectrum and shape after the matched filter.

envelope no longer has a triangular shape and its peak can be approximated by a higher-order polynomial.

As stated in (2.8), the detection range depends on the transmitted energy and is independent of range resolution. Thus, the peak power required to detect a certain object (in the classical pulse radar) is inversely proportional to the range resolution of the pulse radar. Figure 2.13 presents the required transmitted power versus required range resolution, calculated for ideal ($L = 0$) radar operating in the X-band (9.5 GHz), equipped with a 30-dB gain transmitting/receiving antenna and a transmitter enabling the radar to detect a 0-dBsm target at a range of 50 km. It is assumed that the receiver temperature T_R is 300 [K] and detection factor D_o is 13 dB. Note that the presented situation is rather artificial while in almost all pulse radars several pulses are used for target detection (usually 5–40). As a result, the required transmitted power is 5–16 dB smaller than that presented in Figure 2.13, but it is a good illustration of he trade-off between peak power and range resolution in a pulse radar.

The peak power of a transmitted signal in a low-resolution, medium-range radar (resolution 200m, detection range 50 km for 0 dBsm target) must reach the value of 850 kW. If a much higher resolution of 0.2m is required, the transmitted signal power should reach 850 MW. In practice, however, it is very difficult to obtain a peak power greater than 1 MW. The constraints usually come from maximum power densities in radar microwave circuits (ionization in waveguides) and the classical pulse radar resolution for the radar parameters assumed here is limited to several hundred meters. To obtain a better resolution, the pulse compression technique has to be used [2]. In pulse compression radars the transmitted signal duration time is much longer than the time corresponding to the range resolution of the radar. The transmitted signal bandwidth must

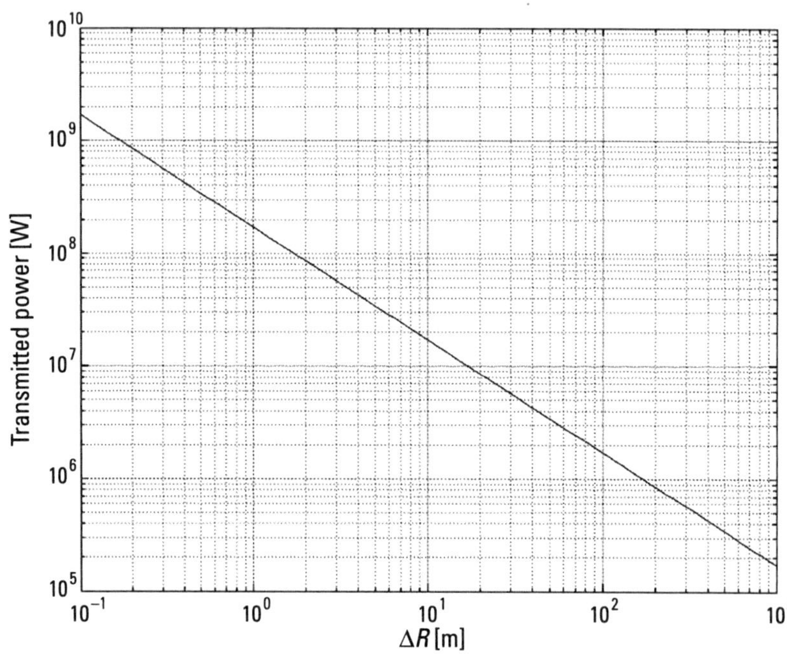

Figure 2.13 Transmitted peak power versus range resolution for single-pulse detection of a 0-dBsm target at a distance of 50 km by radar working in the X-band equipped with a 30-dB gain transmitting/receiving antenna.

be at least $B = 2\Delta R/c$ and the transmitted signal has a more complicated waveform than the simple rectangular pulse filled with a sine signal. In such a case (2.8) held and to obtain a certain detection range the appropriate amount of energy has to be sent towards the target. Thus, the constructors gain an additional dimension of freedom; they can exchange power with time to obtain a constant energy pulse.

In Figure 2.14 the required peak power is plotted versus the pulse time duration (before compression). The radar parameters (band, antenna gain, range) are the same as in the last example.

The use of a 10-μs pulse decreases the required peak power to the level of 100 kW and a 100-μs pulse decreases the peak power to the level of only 10 kW. In the case of a very long (1s) pulse, the required power is equal to 1W. As a rule, a very long pulse allows designers to apply a low peak power solid state transmitter, which is more modern and more reliable than a transmitter equipped with a high-power vacuum tube.

It is worth mentioning that in using pulse compression the radar range resolution is independent of pulse duration and is related to pulse bandwidth. It is also worth mentioning that there are no free lunches in the universe, so we have to pay somehow for such freedom. In the pulse radar using a single

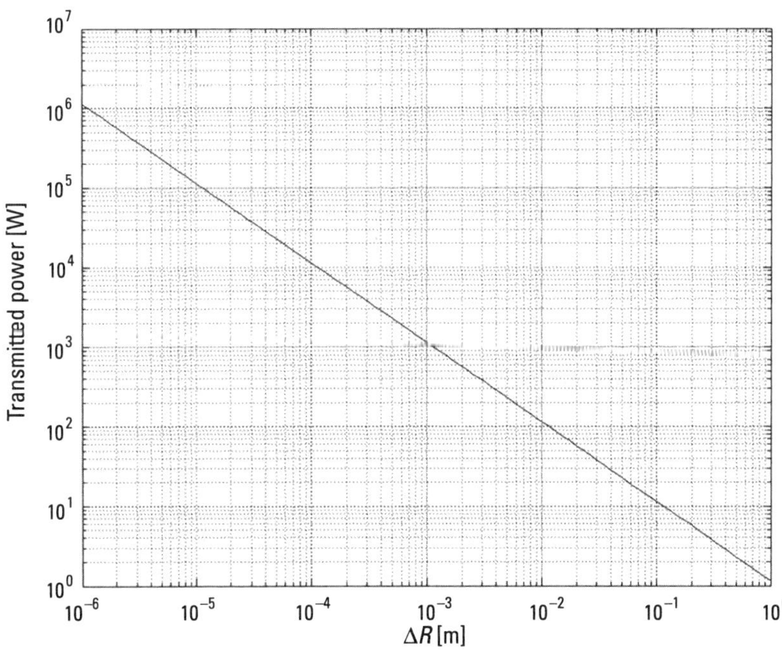

Figure 2.14 Transmitted signal power versus transmitted pulse length for pulse-compression radar working in the X-band equipped with 30-dB gain antenna: a single pulse detection of a 0 dBsm target at the distance of 50 km.

antenna with a transmit/receive switch, we pay when creating a blind zone around the radar: 150m for each microsecond of pulse duration, so typically we can have a 1.5- to 15-km blind zone (for pulses 10–100 μs). If such a blind zone is not acceptable, we can send alternately long pulses to have long-range detection and short pulses to provide detection in the blind zone, but we need extra time to perform this. If longer pulses are used (1 ms–1s, providing a theoretical blind zone of 150–150,000 km), it is necessary to complicate the radar design and have separate transmit and receive antennas with good electromagnetic separation between them.

Another cost of using pulse compression is the existence of time sidelobes (see Figure 2.15), originating from the fact that many targets at different distances can be illuminated simultaneously by the same pulse. The sidelobes of a strong echo can mask weak echoes and weak echoes can be undetected by the radar.

There are a number of waveforms used for pulse compression [2, 4]. The most adequate of them should ensure a narrow peak of the autocorrelation function and low time sidelobes [5–7]. The most frequently used waveform is a chirp modulated signal [8], described by the formula

$$x_T(t) = A(t)e^{2\pi j(Ft + Bt^2/2t_p)} \tag{2.14}$$

for time interval $-t_p/2 < t < t_p/2$, where $A(t)$ denotes the amplitude envelope of the transmitted signal, F denotes the carrier frequency, and B denotes signal bandwidth. In Figure 2.15 the example chirp signal is presented together with its amplitude spectrum and the signal (including sidelobes) at the output of the compression filter. The linear frequency-modulated signal (LFM) has an almost rectangular spectrum. The signal after the matched filtering has a narrow main

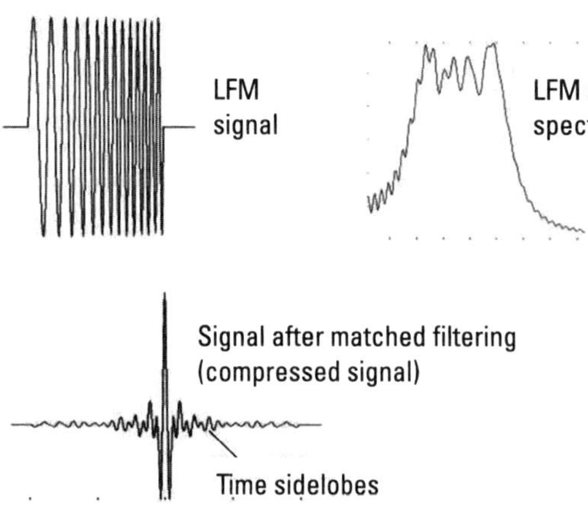

Figure 2.15 Pulse compression using LFM signal with time-bandwidth product = 20.

peak and sidelobes extending over a relatively long time (equal to the transmitted signal duration on each side of the main peak). Because of the rectangular spectrum the time response of the matched filter,for large Bt_p approaches the sinc function. The first sidelobe is at the level of −13 dB below the main peak, the second -19 dB, and the next decrease with reversely proportional to the distance from the main peak as presented in Figure 2.16.

The envelope of an example signal (signal magnitude) at the output of the pulse compression filter for an LFM signal with a time-bandwidth product equal to 20, is presented in Figures 2.16, 2.17, and 2.18. In Figure 2.16 a rectangular amplitude envelope has been used for transmission and the fully matched filter at the receive. The 3-dB length of the main lobe is almost equal to $t_c = 1/B$, but the time sidelobes level is very high (the first sidelobes are 13 dB smaller than the main lobe). In Figure 2.17 the Hamming amplitude envelope (Hamming time window) has been used for tapering the transmission pulse and a fully matched filter was used at the receive side. In this case, the length of the compressed signal main lobe at the 3-dB level is wider than in the previous case and reaches the value of $t_c = 1.8/B$, and the range resolution is almost two times worse than in the previous case. The sidelobes level is now reduced, and the first sidelobes are below -44 dB (compared to the main lobe). A time-bandwidth product close to 20 was used in the first pulse-compression radars, and this value was used in the examples presented above. In modern radars the time-bandwidth product may reach several thousands, but the initial problems remain relevant.

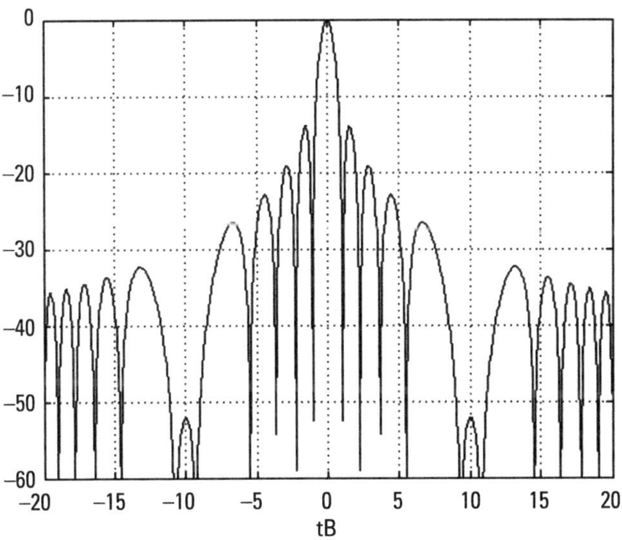

Figure 2.16 The envelope of signal after pulse compression, at the output of the matched filter. Linear frequency modulation, square (boxcar) amplitude of the transmitted signal with time-bandwidth product = 20.

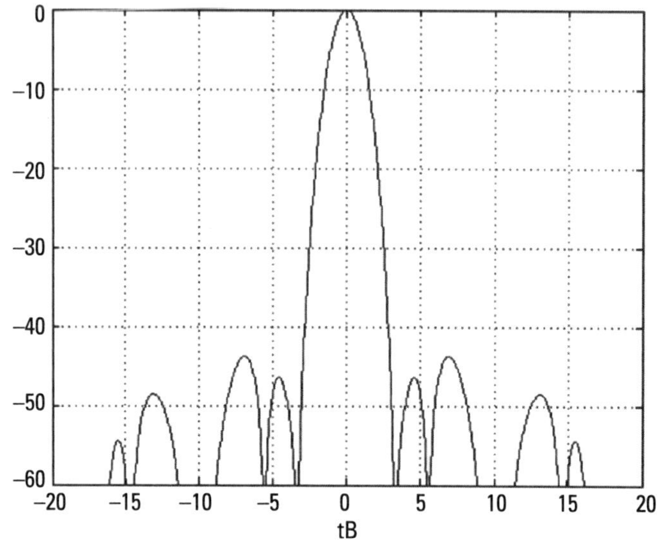

Figure 2.17 The envelope of signal after pulse compression, at the output of the matched filter. Linear frequency modulation, transmitted signal with Hamming window signal with time-bandwidth product = 20.

In many radars it is not possible to modulate the amplitude of the transmitted signal. In such cases the unmatched filter concept is often used. The transmitted signal has a rectangular envelope, but a compression filter is matched to the signal with a Hamming (or other shape) envelope. The signal at the output of the unmatched compression filter with a Hamming window is shown in Figure 2.18. In that case the level of sidelobes is smaller than −29 dB (compared to the main lobe), and the main lobe width is $t_c = 1.4/B$, a little smaller than in the case of a Hamming weighed signal, but wider than for a rectangular (boxcar) weighted signal. The range resolution is 1.4 times worse than in the case when no windowing is applied. Generally, the sidelobe level and structure for a windowed signal depend not only on the shape of the applied window but also on the time-bandwidth product of the transmitted signal.

Linear frequency modulation is used extensively in both radar technology and in nature. These sounding signals are used by bats, dolphins, and other animals. The modification of linear frequency modulation results in nonlinear frequency modulation, where, instead of applying time windows, modulation nonlinearity is used [4]. The number of possible linear modulations for a given pulse time duration and bandwidth is limited to two: up-chirp and down-chirp. They are almost orthogonal. When nonlinear frequency modulation is used, there are a few more possibilities, but still only two are almost orthogonal, while others are strongly correlated. In some applications a greater variety of sounding pulses are required. In such cases phase modulation is often used. The

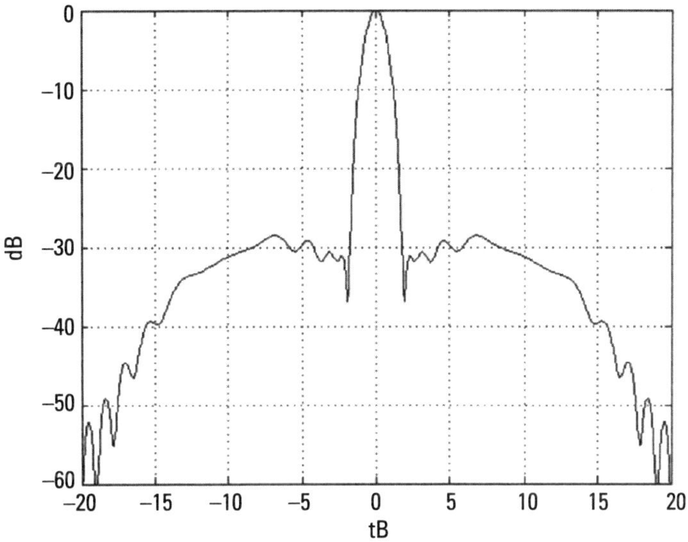

Figure 2.18 The envelope of signal after pulse compression, at the output of the unmatched filter. Linear frequency modulation, rectangular (boxcar) amplitude of transmitted signal, Hamming window used at the receiving side.

simplest are binary phase (biphase) modulation [9, 10] or polyphase waveforms [11–13]. Among others, the binary-phase Barker code is also often used [14, 15]. The amplitude of the main peak of the correlation function of the Barker code is equal tothe length of the code, while that of the sidelobes is limited to unity. There are Barker codes that exist up to and including length of 13. The absolute value of the correlation function of a widely used 13-stage code (1111100110101) is presented in Figure 2.19. As can be seen, the time sidelobes are uniformly distributed at the level of −22 dB.

The duration of the transmitted pulse is constrained by several technical factors, depending on the radar function performed and the selected processing schema. In classical surveillance scanning radar without pulse compression it is limited by the single direction scanning time described by (2.9). In all radars using classical signal processing (i.e., without compensation for target velocity), the duration of the pulse is also limited by the time the target stays in the range-azimuth resolution cell. The time that targets remain in a single-range resolution cell is approximated by the formula

$$t_p < \frac{\Delta r}{v_{\max}} = \frac{c}{2Bv_{\max}} \tag{2.15}$$

where $v_{\max}$ denotes the maximum target radial velocity. For a medium-resolution radar ($\Delta r = 50$m), and assumption $v_{\max}$ = Mach 3 (approximately 1,000

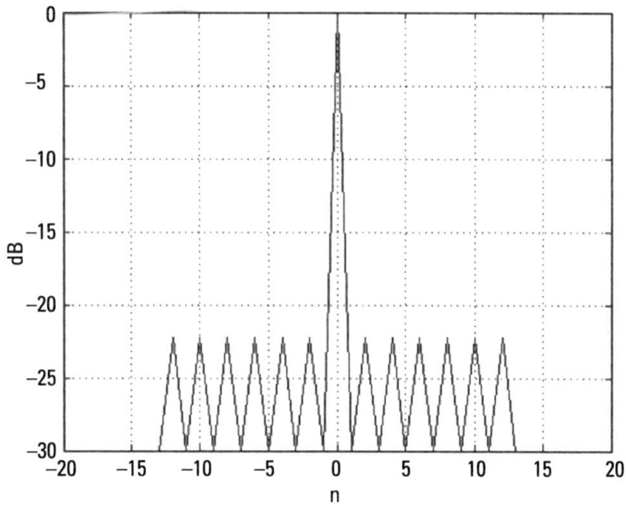

Figure 2.19 The magnitude of signal after pulse compression, at the output of the matched filter; 13-point Barker (binary phase) code.

m/s), the maximum pulse duration time has to be smaller than 50 ms, unless the signal processor includes a correction for known target velocity. The third factor is the time when the target remains inside the antenna beam (for target movement perpendicular to the line of sight). This time limit can be expressed as follows:

$$t_p < \frac{\alpha_{ant} R}{v_{\max}} \tag{2.16}$$

where R denotes the distance to the target and α_{ant} denotes the antenna beamwidth (plane angle in radians, for a fixed beam angle). For a 20-mrad antenna beamwidth, a range of 10 km, and a target velocity of 1,000m/s, the pulse duration time has to be shorter than 200 ms. For most radars this constraint is less restrictive than the previous ones. Using more advanced signal processing (like signal stretching to incorporate range migration or multibeam coherent processing to prevent angular migration), it is possible to extend the pulse duration or coherent integration time.

2.3 Radar Range Velocity Measurement and Range Velocity Resolution

Classical radar can measure not only the distance to the target but also the target radial velocity by exploiting the Doppler effect. The returning (received) signal is a delayed copy of the transmitted one only when the signal is reflected from a stationary (nonmoving) target. In most cases, however, the radar is used for the detection of moving targets. Consequently, the return signal, reflected from a moving target, can be described by

$$x_R^{HF}(t) = A(R(t))x_T^{HF}\left(t - \frac{2R(t)}{c}\right) \tag{2.17}$$

where $x_T^{HF}(t)$ denotes the transmitted (usually high frequency) signal, $x_R^{HF}(t)$ denotes the received signal, $R(t)$ denotes the distance between the radar and the target, and $A(t)$ denotes the amplitude factor, which can be calculated on the basis of (2.4). Let us consider a narrowband transmitted signal

$$x_T^{HF}(t) = x_T(t)\exp(j2\pi Ft + j\phi) \tag{2.18}$$

where $x_T(t)$ is the complex envelope of the signal, F is the carrier frequency, and φ is the starting (initial) phase. The received signal can be approximated by the formula

$$x_R^{HF}(t) = A\left(r(t)\right)x_T\left(t - \frac{2R(t)}{c}\right)\exp\left(j2\pi\left(Ft - \frac{2R(t)F}{c}\right) + j\phi\right) \tag{2.19}$$

If we assume that the target moves with constant radial velocity v, then the radial radar-target distance changes linearly with respect to time:

$$R(t) = R_o + vt \tag{2.20}$$

In this situation the received signal can be expressed as

$$\begin{aligned} x_R^{HF}(t) &= A\left(r(t)\right)x_T\left(t - \frac{2(R_0 + vt)}{c}\right)\exp\left(j2\pi\left(F - \frac{2vF}{c}\right)t + j\phi_R\right) = \\ &A\left(r(t)\right)x_T\left(t\left(1 - \frac{v}{c}\right) - \frac{2R_0}{c}\right)\exp\left(j2\pi(F - f_d)t + j\phi_R\right) \end{aligned} \tag{2.21}$$

where $\varphi_R = \varphi - 4\pi R_0 F/c$ is the received signal initial phase, and $f_d = 4vF/c = 2v/\lambda$ is a Doppler frequency shift. For radar using a short single pulse, when the Doppler shift is much smaller than the reciprocal of the transmission pulse interval ($4vF/v << 1/t_T$), the Doppler effect is negligible. In pulse radar it is possible to estimate the target Doppler shift if the train of pulses is used. According to (2.21), the phase of each returned pulse is different, and Doppler information is hidden in their phase difference. Table 2.3 presents typical Doppler frequencies for different targets and different radar bands.

The emission of a consecutive pulse train can be treated as a time-sampling procedure. According to the Shannon sampling theorem, the pulse repetition frequency (PRF) should be twice the maximum Doppler frequency:

$$PRF > \frac{4v_{\max}}{\lambda} \tag{2.22}$$

For example, for a 10-cm wavelength (3 GHz) and a 1,000-m/s maximum target velocity, the required PRF should not be smaller than 40 kHz and the target velocity can be estimated finding the coordinates of the maximum FFT of the consecutive echoes sampled for the chosen time delay (range to the target).

Such a relatively high PRF limits the radar unambiguity range to the value

$$R_{\max} = \frac{c}{2PRF} = \frac{c\lambda}{8v_{\max}} \tag{2.23}$$

In the above example, the unambiguous range is limited to 3,750m. The typical detection range for medium-range radar is 100 km, so in most pulse radars it is not possible to simultaneously achieve range and the velocity unambiguity. In a ground-based surveillance radar range unambiguity is usually treated as a major requirement. The pulse repetition frequency must be selected at a value far below the constraint formulated in (2.22), and in effect the velocity measurement is ambiguous. Using the nonuniform sampling concept, a number of methods for unambiguous velocity measurement have been developed [16], but low-frequency components (echoes originating from the

Table 2.3

Doppler Frequencies for Different Targets and Different Radar Bands

Target Velocity [m/s]	Doppler Frequency [kHz] for Radar Working at Frequency/Band: 300 MHz P	1.7 GHz L	3.3 GHz S	5.2 GHz C	9.4 GHz X	24 GHz K	94 GHz W	Target Example
1	0.002	0.011	0.022	0.034	0.062	0.160	0.626	Walking man
10	0.020	0.113	0.220	0.346	0.626	1.600	6.266	Bike
30	0.060	0.340	0.660	1.040	1.880	4.800	18.800	Car
100	0.200	1.133	2.200	3.466	6.266	16.000	62.666	Small plane
300	0.600	3.400	6.600	10.400	18.800	48.000	188.000	Jet plane
600	1.200	6.800	13.200	20.800	37.600	96.000	376.000	Jet fighter
1,200	2.4000	13.600	26.400	41.600	75.200	192.000	752.000	Rocket

ground, vegetation, weather, and sea clutter) can mask weak, high-frequency echoes. For airborne radars [16–18], medium [19] or high PRF concepts [20, 21] are usually applied.

The high PRF mode is a radar sounding mode in which the PRF is selected in such a way that velocity measurement is unambiguous. This means that constraint (2.22) is fulfilled. For the L-band radar the PRF should be around 20 kHz, for the X-band 100 kHz, and for the K-band 200 kHz. Of course, in that mode velocity measurement is unambiguous, while range measurement is ambiguous, and a special procedure, usually based on the Chinese remainder theorem, has to be used to resolve range ambiguity [23]. The radial velocity resolution is limited by the time on target. Using the filtering concept or equivalent FFT processing, it is possible to separate two targets in velocity when the Doppler frequency difference is greater than the reciprocal of a coherent integration time (time duration of a signal block taken for Doppler frequency analysis, smaller or equal to time on target). The velocity resolution is as follows:

$$\Delta v = \frac{2\lambda}{t_i} = \frac{2c}{t_i F} \tag{2.24}$$

For example, for a 10-cm wavelength and an integration time of 10 ms, the velocity resolution is 20 m/s, while for 1s the resolution is much better and equal to 0.2 m/s. The coherent integration time is limited by the factors mentioned earlier, among which is the time when a target remains in the range resolution cell (2.15). Additionally, when Doppler processing is applied, the integration time is limited by target maneuver. Proper target detection and velocity estimation can be achieved only in the situation where the target velocity during observation remains in a velocity resolution cell. This leads to the constraint

$$t_i < \sqrt{\frac{2\lambda}{a_{\max}}} \tag{2.25}$$

where $a_{\max}$ denotes the maximum target range acceleration. For example, for a 10-cm wavelength with 1-g (9.81 m/s) maximum target acceleration, the coherent integration time is limited to 0.14s, while for 10g the limitation is at 0.045s. It is worth noting that target range acceleration can be present even for nonmaneuvering targets, while range and range velocity are measured naturally in radar polar coordinate systems. Signal processors can be designed to overcome this limitation by applying compensation for known or assumed target acceleration, requiring multiple channels if the acceleration is unknown.

2.4 Radar Ambiguity Function

Up to now, the range and Doppler measurements have been treated as independent measurements. It is true for pulse radar, when short pulses are used, but it is not fully true for pulse compression radars. To measure and verify coupling between range and Doppler measurements, the radar ambiguity function is used [23–28].

The radar ambiguity function is a 2-D function of the Doppler frequency and the range (or time delay). Since Doppler frequency and range (time delay) are canonical conjugate variables, there exists an uncertain relationship between them. It is not possible to measure both these quantities with infinite accuracy. The shape of the region of uncertainty depends on the transmitted waveform. The cross-ambiguity function between the two signals x_1 and x_2 is defined by the following formula:

$$c(\tau,\phi) = \int x_1\left(t - \frac{\tau}{2}\right)x_2^*\left(t + \frac{\tau}{2}\right)e^{j\phi t}\,dt \tag{2.26}$$

The independent variable τ represents the time shift between two signals, while the variable φ represents the shift of angular frequency between signals. The ambiguity function differs from what was described in (2.26); it shows time-Doppler properties of a single signal, and has the form

$$a(\tau,\phi) = \int x_1\left(t - \frac{\tau}{2}\right)x_1^*\left(t + \frac{\tau}{2}\right)e^{j\phi t}\,dt \tag{2.27}$$

Examples of ambiguity functions of signals commonly used in radar technology are presented in Figures 2.20 through 2.24. In all these cases it is assumed that a radar working in the X-band is used (10-GHz carrier frequency).

In Figure 2.20 the absolute value of the ambiguity function of the short (100-ns), single-frequency pulse is presented. As follows from the figure, the time (range) resolution is good (100 ns is equivalent to 15-mach resolution), but there is almost no Doppler resolution using single pulse (within the range ±70 kHz, which is equivalent to a target radial velocity ±3 Mach.)

In Figure 2.21 the absolute value of the ambiguity function of the continuous-wave, single-frequency radar is presented. The frequency resolution, for the 0.1s analysis time, is equal to 10 Hz, and the velocity resolution is equal to 0.15 m/s, respectively. There is no range resolution. Such radars commonly used for velocity measurements include single police gun radars, radars for measuring tennis and golf ball velocity, and so forth.

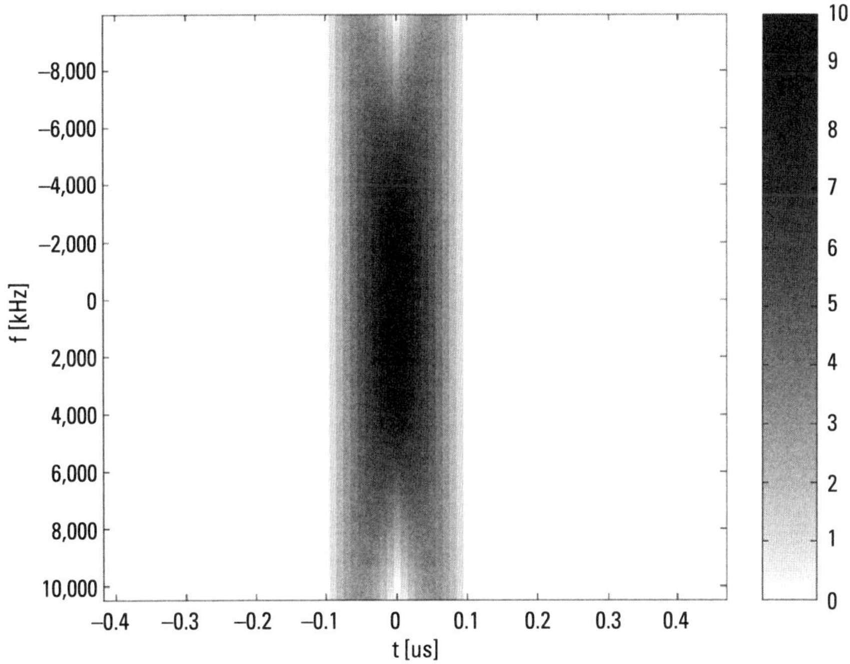

Figure 2.20 The absolute value of ambiguity function for a short transmitted pulse. The function magnitude is coded in grayscale.

The absolute value of the ambiguity function of a 10-μs chirp signal (10-MHz bandwidth) is presented in Figure 2.22. For this type of waveform, the range resolution is very good (15m in the example presented). The Doppler frequency changes the time position of the signal at the output of the matched filter. However, in the useful Doppler frequency range correspondence ±2M to ach, the time shift is limited to two range resolution cells. A relatively high Doppler frequency shift decreased the amplitude of the compressed signal, but again a Doppler shift corresponding to the maximum target velocity caused only very small amplitude changes (below 1 dB). However, if much longer pulses are used, then the Doppler frequency ranges will be smaller and range shift due to the presence of target movement will be greater. In this case range correction should be applied after the estimation of Doppler frequency.

In Figure 2.23 the absolute value of the ambiguity function of the 13-bit Barker code (100-ns bit rate) is presented. In this case the range resolution is also good (as in the previous case). There is no range displacement related to the Doppler shift, but the useful Doppler velocity range is much smaller. The faster target will not only be attenuated but also split in range, so such a sounding signal is useful only for relatively slow targets unless velocity compensation is applied in the processor.

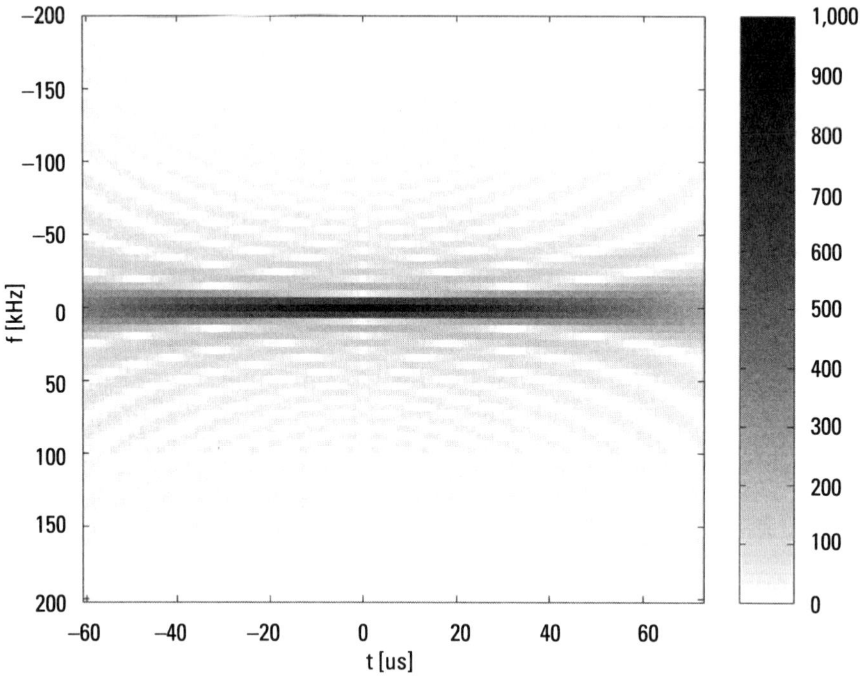

Figure 2.21 The absolute value of the ambiguity function for a single-frequency continuous wave. The function magnitude is coded in grayscale.

The absolute value of the ambiguity function calculated for a long noise pulse (1-ms duration time, 10-MHz bandwidth, X-band) is presented in Figure 2.24. Due to the long pulse time, the main peak of the ambiguity function is narrow not only in range dimension, but also in the Doppler frequency dimension. The useful Doppler frequency range is much wider than the ambiguity function main lobe. This means that both the range and Doppler frequency can be determined using this kind of modulation, and it is possible to distinguish the two targets at the same distance to the radar, but having different radial velocities. The disadvantage of this kind of modulation is that in using a long noise pulse it is necessary to apply a bank of matched filters adopted to different Doppler frequencies instead of a single matched filter. Detailed analyses of the applications of noise waveforms in different radars are the subject of the next chapters.

2.5 Radar Operation over the Ground or the Sea

In the previous sections all considerations have been made assuming a single target and free-space conditions. In practice a radar is situated either on the

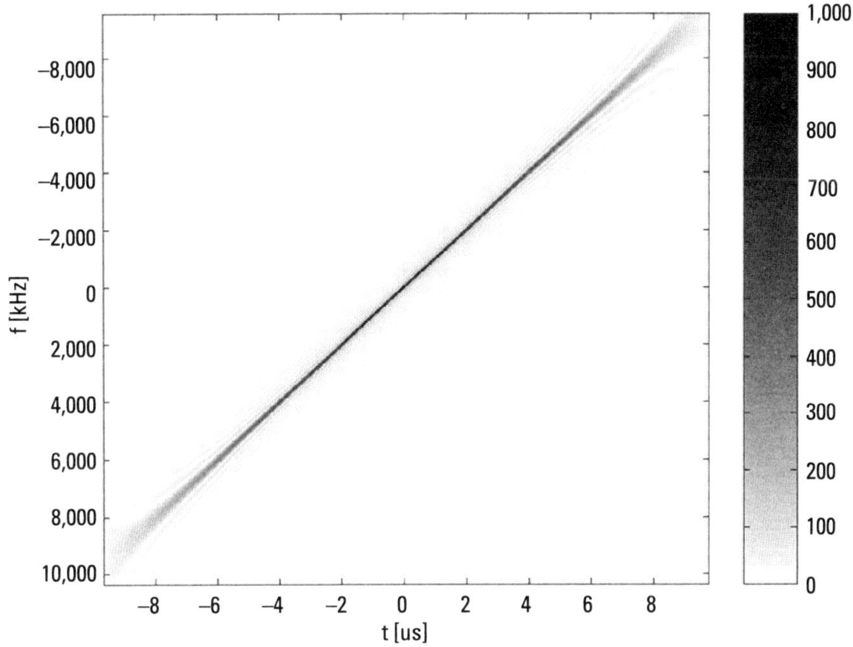

Figure 2.22 The absolute value of the ambiguity function of the chirp signal. The function magnitude is coded in grayscale.

ground or on a moving platform such as a vessel, airplane, or spacecraft. If the radar uses a pencil beam and directs the beam towards the sky, then the free-space condition is a good approximation. If the transmit or receive beam looks towards the ground or sea surface, then such an approximation is very coarse and to understand all phenomena related to the radar, it is necessary to also take into consideration the reflection from the ground or sea surface.

2.6 Multipath Effect

A target located over the ground or sea surface is illuminated in two ways: by a direct wave and by an indirect wave reflected by the ground surface. A typical scenario over flat ground is presented in Figure 2.25. To calculate the effect of the multiple paths, it is necessary to calculate the complex signals traveling in both ways and sum them.

The direct path length is described by

$$R_1 = \sqrt{L^2 + (h_T - h_t)^2} \tag{2.28}$$

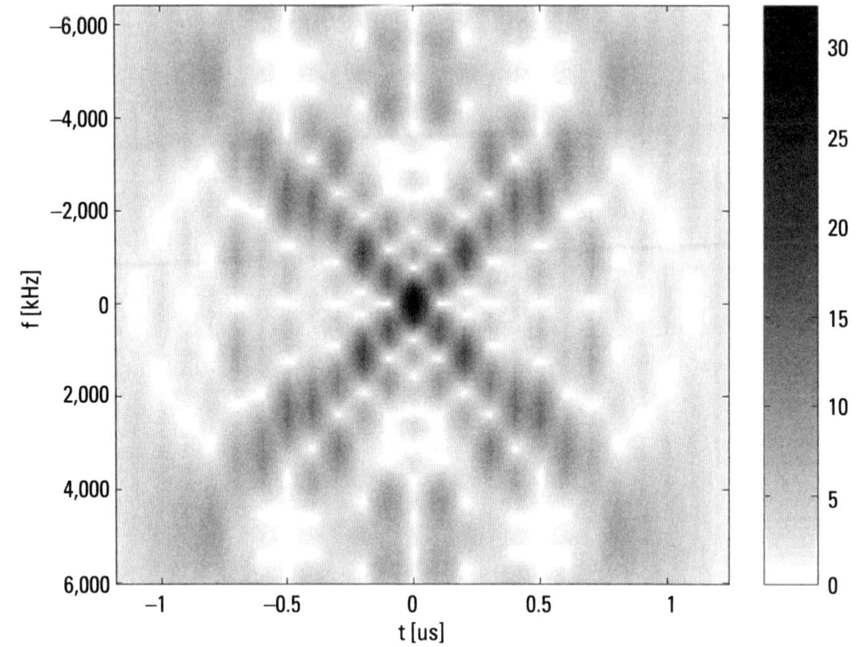

Figure 2.23 The absolute value of ambiguity function of 13-bit Barker code signal. The function magnitude is coded in grayscale (linear coding).

where L is the ground distance between the transmitter and the target, h_T is the altitude of the target, and h_t is the height (over ground level) of the transmitting antenna.

The indirect path length can be easily calculated for specular reflection from a flat Earth as presented in Figure 2.25. This length is described by

$$R_1 = \sqrt{L^2 + (h_T + h_t)^2} \tag{2.29}$$

The difference between those two distances is equal to

$$\Delta R = \sqrt{L^2 + (h_T + h_t)^2} - \sqrt{L^2 + (h_T + h_t)^2} \tag{2.30}$$

If ΔR is significantly smaller than the radar range resolution, described by (2.12), then it is possible to assume that the illuminating baseband signal X_I is modified only in amplitude and phase according to the following equation:

$$X_I = X\left(1 + Ae^{-\frac{2\pi\Delta R}{\lambda}}\right) \tag{2.31}$$

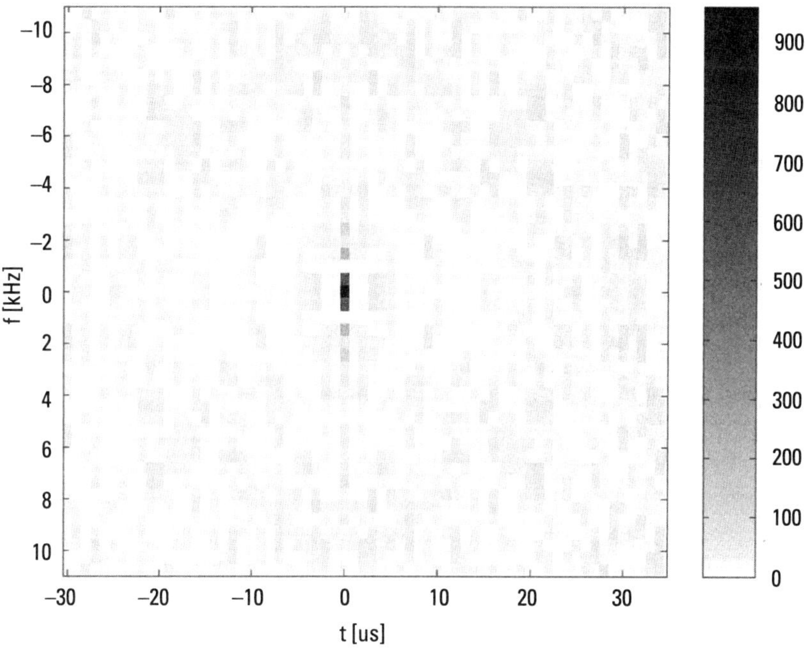

Figure 2.24 The absolute value of the ambiguity function of a noise signal. The function magnitude is coded in grayscale.

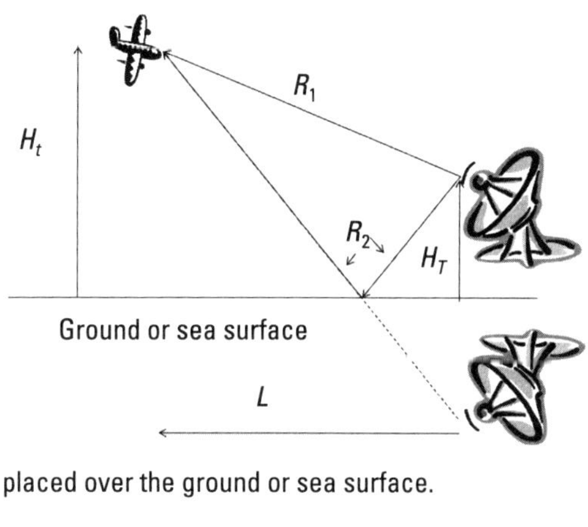

Figure 2.25 Radar placed over the ground or sea surface.

where X is the baseband complex transmitted signal, A is the ground reflection coefficient (usually negative, where the ground reflection reverses the signal phase), and λ is the wavelength of the illuminating signal.

In the classical approach, described by (2.31), the time shift between the direct and indirect signals is not taken into consideration. When the distance difference is comparable or higher than the radar range resolution, a more complex analysis has to be performed, including correlation between pulses coming both ways. A large distance between the two paths leads to reverses of the direct and indirect signals. When the distance difference is close to the width of the range resolution cell, the correlation coefficient is almost equal to 0, and instead of the signal fading due to opposite phase summation, one can observe an increase of total illuminating energy by 3 dB. As a side effect, degradation of range resolution can be observed. If the range distance (2.30) is significantly greater than the radar range resolution, then two targets can be observed instead of one. The true one, usually stronger, is closer to the radar, while the second is placed at a longer distance.

For a rectangular spectrum signal the correlation function has the form of a *sinc* function. For a Gaussian signal spectrum, the correlation function is also Gaussian. In the case of a strong correlation of the direct and indirect signal, the illuminating power can vary from almost 0 to double that of the signal amplitude (6 dB). The lower and upper bounds for the illumination power change caused by the multipath effect versus the distance difference are plotted in Figure 2.26 for the rectangular signal spectrum and in Figure 2.27 for the Gaussian signal spectrum.

As we can see, the bounds depend very slightly on the shape of the spectrum of the illuminating signal if the range shift is smaller than the range

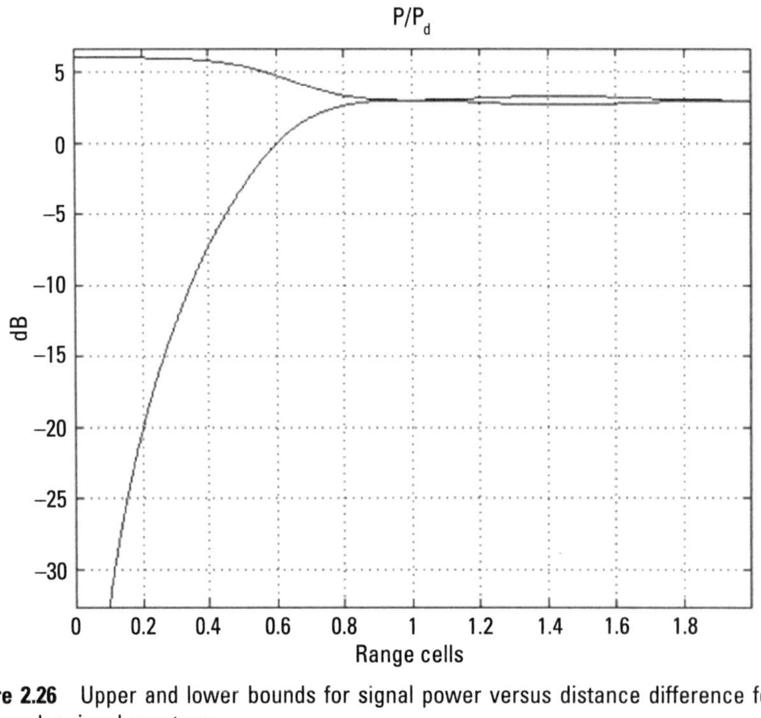

Figure 2.26 Upper and lower bounds for signal power versus distance difference for the rectangular signal spectrum.

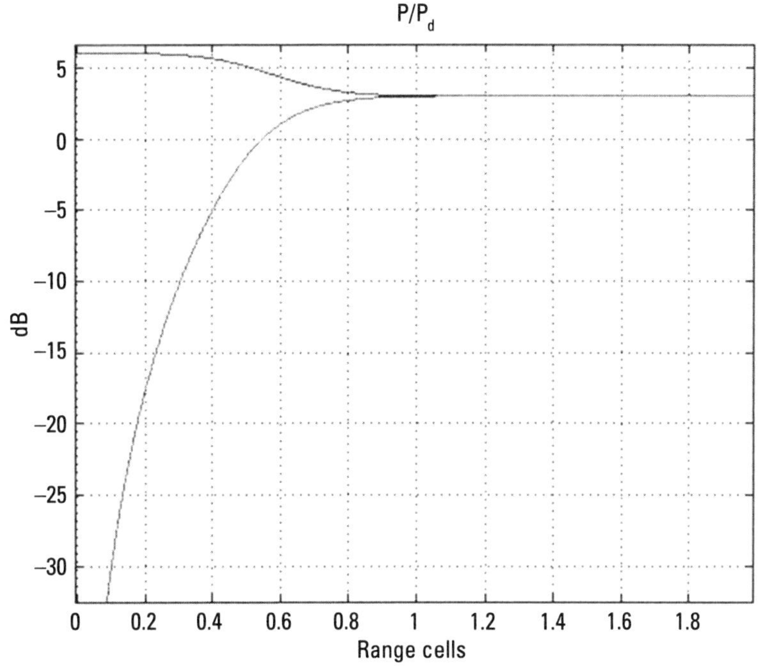

Figure 2.27 Upper and lower bounds for signal power versus distance difference for the Gaussian signal spectrum.

resolution cell. For the Gaussian spectrum there is almost no illumination power change if the range distance is within the second resolution cell. For rectangular illumination the power changes are small, within 1 dB.

In the case when ΔR is bigger than two range resolution cells instead of summing energy, it is necessary to consider multiple detections of different ranges and different radial velocities.

A similar analyses can be performed for propagation from the target to the receiver. Thus, four different ways can be observed:

- Direct illumination–direct reception;
- Indirect illumination–direct reception;
- Direct illumination–indirect reception;
- Indirect illumination–indirect reception.

If the radar receiver is in the same place as the radar transmitter (monostatic case), then either the signal variation is twice as high (in decibels) or three targets are visible (direct path, direct-indirect path, and indirect-indirect path). In the multistatic radar up to four targets can be visible.

The flat Earth model is adequate only for small distances between the target, transmitter, and receiver. For longer distances it is necessary to apply a spherical Earth model. The results of the calculation of the distance difference obtained from the two models for an object at altitudes of 1 and 10 km, a radial velocity of 100 m/s, and placing the radar transmitter at 200m AGL, are presented in Figure 2.28. As can be clearly seen, the flat Earth model can be used for short distances (up to 10 to 20 km), while for longer distances the spherical model should be used. It is worth noting that at higher altitudes there are very small differences between the flat and spherical Earth models.

To obtain multiple targets, relatively flat (compared to wavelength) areas are required around the radar transmitters. The change of the illuminator-target distance also influences the Doppler frequency shift. While direct and indirect illumination have different path lengths and illuminate the target at different angles, in addition to the difference in range target position, it is possible to observe the difference in target radial velocity. The radial velocity differences for the conditions described above are presented in Figure 2.29.

The velocity differences are rather small and in most cases do not exceed 1 m/s (for 100 m/s target velocity). The velocity resolution depends on integration time and wavelength, but in most cases it will not be possible to distinguish between direct path and indirect path echo in the velocity dimension.

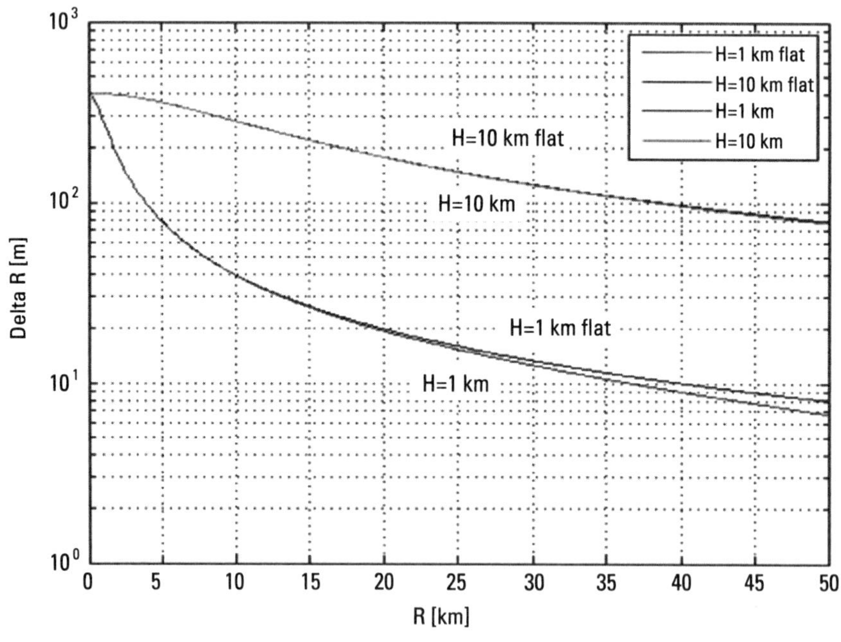

Figure 2.28 The length difference in direct and indirect path for radar transmitter height 200m for flat and spherical Earth models.

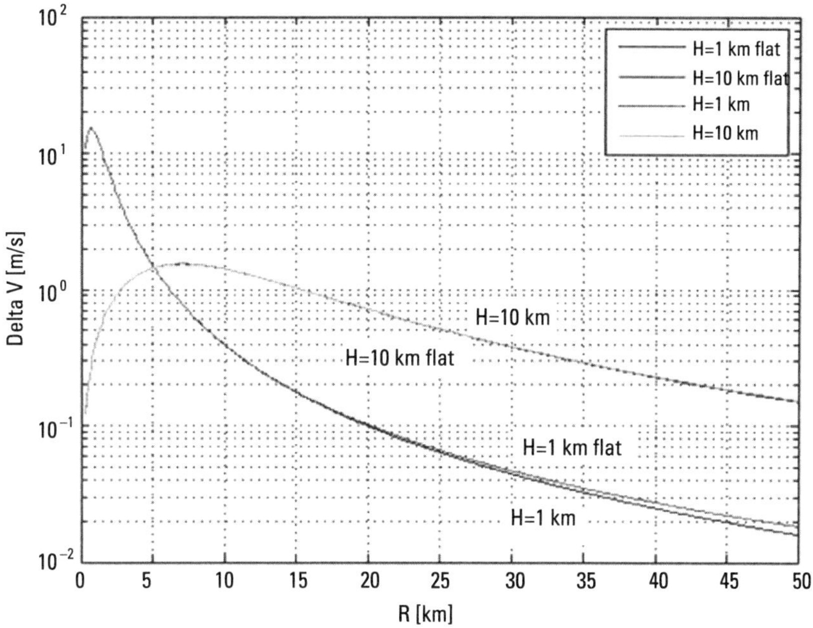

Figure 2.29 The radial velocity difference in direct and indirect path for transmitter height 200m, target radial speed 100 m/s, and flat and spherical Earth models.

In this chapter the fundamentals of classical radar technology were presented to provide to the reader with basic knowledge. In the following chapters advanced techniques used in noise radar will be presented. The radar fundamentals presented are dated from the beginning of the second half of the twentieth century. Since that time, progress in pulse radar has been amazing and a great deal of modern signal processing technology has been developed. Readers can find information on modern radar signal processing in [2, 29, 30] and in many other books and journal papers.

References

[1] Hülsmeyer, C., Verfarhen, um entfernte metalishe Gegenstandemittels electrishe Wallen inem Beobachter zu Melden, German Patent No. 165546, issued April 30, 1904.

[2] Skolnik, M. I., *Radar Handbook*, 2nd ed., New York: McGraw-Hill, 1990.

[3] McDonough, R. N., and A. D. Whalen, *Detection of Signals in Noise*, 2nd ed., New York: Academic Press, 1995.

[4] Cook, C. E., and M. Bernfeld, *Radar Signals: An Introduction to Theory and Application*, New York: Academic Press, 1967; Norwood, MA: Artech House, 1993.

[5] Judd, G. W., "Technique for Realizing Low Time Sidelobe Levels in Small Compression Ratio Chirp Waveforms," *Proc. IEEE Ultrasonics Symp.*, 1973, pp. 478–481.

[6] Temes, C. L., "Sidelobe Suppression in a Range Channel Pulse-Compression Radar," *IRE Trans* Vol. MIL-6, April 1962, pp. 162–169.

[7] Turyn, R., and J. Stover, "On Binary Sequences," *Proc. Am. Math. Soc.,* Vol. 12, June 1961, pp. 394–399.

[8] Klauder, J. R., et al., "The Theory and Design of Chirp Radars," *Bell Syst. Tech. J.*, Vol. 39, July 1960, pp. 745–808.

[9] Lindner, J., "Binary Sequences Up to Length 40 with Best Possible Autocorrelation Function," *Electron. Lett.,* Vol. 11, October 1975, p. 507.

[10] Golomb, S. W., and R. A. Scholtz, "Generalized Barker Sequences," *IEEE Trans.,* Vol. IT-11, October 1965, pp. 533–537.

[11] Frank, R. L., "Polyphase Codes with Good Nonperiodic Correlation Properties," *IEEE Trans.*,Vol. IT-9, January 1963, pp. 43–45.

[12] Lewis, B. L., and F. F. Kretschmer, Jr., "A New Class of Polyphase Pulse Compression Codes and Techniques," *IEEE Trans.,* Vol. AES-17, May 1981, pp. 364–372 (correction, *IEEE Trans.,* Vol. AES-17, May 1981, p. 726).

[13] Luenburger, D. G., "On Barker Codes of Even Length," *Proc. IEEE,* Vol. 51, January 1963, pp. 230–231.

[14] Turyn, R., "On Barker Codes of Even Length," *Proc. IEEE* (correspondence), Vol. 51, September 1963, p. 1256.

[15] Schleher, D. C., (ed.), *MTI Radar*, Dedham, MA: Artech House, 1978.

[16] Dickey, F. R., Jr., "Theoretical Performance of Airborne Moving Target Indicators," *IRE Transactions on Airborne Electronics,* Vol. PGAE-8, June 1953, pp. 12–23.

[17] Perkins, L. C., H. B. Smith, and D. H. Mooney, "The Development of Airborne Pulse Doppler Radar," *IEEE Trans.,* Vol. AES-20, May 1984, pp. 290–303.

[18] Clarke, J., "Airborne Early Warning Radar," *Proc. IEEE,* Vol. 73, February 1985, pp. 312–324.

[19] Hovanessian, S. A., "Medium PRF Performance Analysis," *IEEE Trans. Aerosp. Electron. Syst.*, Vol. AES-18, May 1982, pp. 286–296.

[20] Skillman, W. A., and D. H. Mooney, "Multiple High-PRF Ranging," *Proc. IRE Conf. Mil. Electron.,* 1961, pp. 37–40. Reprinted in Barton, D. K., *CW and Doppler Radars,* Vol. 7, Dedham, MA: Artech House, 1978, sec. IV-I, pp. 205–213.

[21] Hovanessian, S. A., "An Algorithm for Calculation of Range in Multiple PRF Radar," *IEEE Transactions on Aerospace and Electronic Systems,* Vol. AES-12, March 1976, pp. 287–289.

[22] Rendas, M. J. D., and J. M. F. Moura, "Ambiguity in Radar and Sonar," *IEEE Transactions on Signal Processing,* Vol. 46, No. 2, February 1998, pp. 294–305.

[23] Sinsky, A. I., and C. P. Wang, "Standardization of the Definition of the Radar Ambiguity Function," *IEEE Transactions on Aerospace and Electronic Systems*, Vol. AES-10, No. 4, July 1974, pp. 532–533.

[24] Hussain, M. G. M., "Ambiguity Functions for Monostatic and Bistatic Radar Systems Using UWB Throb Signal," *IEEE Transactions on Aerospace and Electronic Systems*, Vol. 47, No. 3, July 2011, pp. 1710–1722.

[25] San Antonio, G., D. R. Fuhrmann, and F. C. Robey, "MIMO Radar Ambiguity Functions," *IEEE Journal of Selected Topics in Signal Processing*, Vol. 1, No. 1, June 2007, pp. 167–177.

[26] Tsao, T., et al., "Ambiguity Function for a Bistatic Radar," *IEEE Transactions on Aerospace and Electronic Systems*, Vol. 33, No. 3, July 1997, pp. 1041–1051.

[27] Cohen, S. A., "Cross-Ambiguity Function for a Linear FM Pulse Compression Radar," *IEEE Transactions on Electromagnetic Compatibility*, , Vol. EMC-14, No. 3, August 1972, pp. 85–91.

[28] Levanon, N., and A. Freedman, "Ambiguity Function of Quadriphase Coded Radar Pulse," *IEEE Transactions on Aerospace and Electronic Systems*, Vol. 25, No. 6, November 1989, pp. 848–853.

[29] Franceschetti, G., and R. Lanari, *Synthetic Aperture Radar Processing*, Boca Raton, FL: CRC Press, 1999.

[30] Richards, M. A., J. A. Scheer, and W. A. Holm, *Principles of Modern Radar: Basic Principles*, Raleigh, NC: SciTech Publishing, 2010.

3

Noise Radar

In the previous chapter the fundamentals of radar technology were shown. The different sounding waveforms were discussed starting from simple short pulses, continuing on to linear frequency modulation, and through to more complicated phase coded signals. In this chapter sounding signals in the form of random or pseudo-random sequences will be the focus of consideration. The properties of such sounding signals will be discussed and the method of processing signals to obtain good detection properties will be shown.

The term "noise radar" refers to a group of radars using a random or pseudo-random waveform signal for target illumination. In many references, this type of radar is called a random signal radar (RSR) [1]. This type of radar can be used in a relatively wide range of applications. It is possible to construct surveillance, tracking, guidance, collision warning, subsurface, and other radars using noise-sounding signals. Noise radars have several advantages over the classical pulse, pulse-Doppler, and FMCW radars. From the literature it follows that a noise waveform signal ensures the absence of range and Doppler ambiguity, low peak power, and very good electromagnetic compatibility with other devices sharing the same frequency band. Due to the low peak power, noise radars also have a very good electronic countercountermeasure (ECCM) capability. The probability of the interception of this kind of radar is also very low, and even if the sounding signal is detected, it is difficult to classify and treat it as the threat.

Unfortunately, this type of radar also has several disadvantages. Undoubtedly, signal processing in a noise radar is much more complicated in comparison to the corresponding pulse radar. The required computational power is very high and it is difficult to apply the noise radar signal processing in real time for high-bandwidth radars. Noise radars also suffer from near-far problems, very well known in radio communication. The received electromagnetic power changes with the reciprocal of the fourth power of the range, so for long-range radars a very high, effective dynamic range (usually much higher than 100 dB) is required. For a smaller effective dynamic range, the masking effect will be clearly visible, and strong objects nearby will mask weak ones at a far distance.

The first fundamental paper on noise radar was published by B. M. Horton in 1959 [2], presenting the major concept of a range measuring radar. The next papers in this area were published in 1960s and 1970s [3, 4]. At that time, the concept of noise radar did not attract many radar designers since the correlation signal processing was very difficult to obtain by means of analog electronic circuits. However, several researchers had been working on the noise radar concept. Two of them were M. P. Grant and G. R. Cooper from Purdue University [5]. They proposed the use of the analog delay line to compensate for the time delay of the echo signal and the use of an analog multiplier. The block diagram of their radar is presented in Figure 3.1.

Such a radar can only detect slow targets (with the Doppler frequency lower than the bandwidth of a lowpass filter) at the single range gate selected by the delay of the applied delay line.

Of course, it was possible to detect targets at different ranges by using a multiple receiving channel, as presented in Figure 3.2. By selecting the appropriate delays of an applied delay line, it is possible to select the range cells for which detections are performed.

Another approach to noise radars was proposed by P. R. Krehbiel and M. Brook in 1979 [6]. They used a pulse noise radar working in the X-band (λ = 3 cm) with a pulse width of 1 μs and a bandwidth B of 300 MHz. The noise signal was generated by amplifying resistor thermal noise using a band limiting amplifier. The output signal was keyed to form 1-μs pulses sent by the Tx/Rx antenna. The received signal, separated from the transmitted signal by the

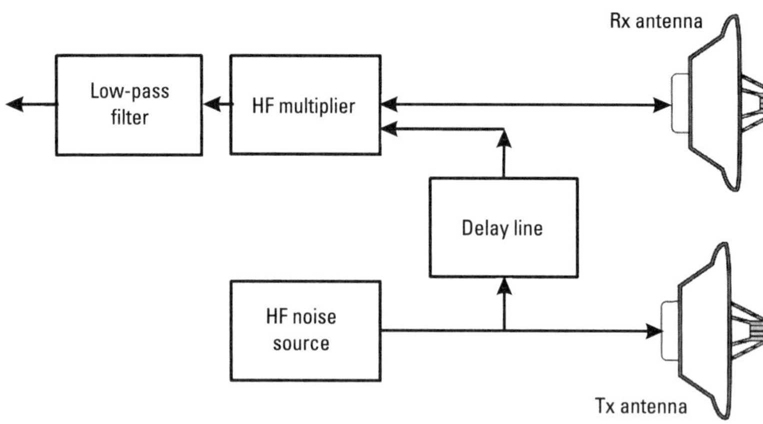

Figure 3.1 Block diagram of the noise radar based on the delay line.

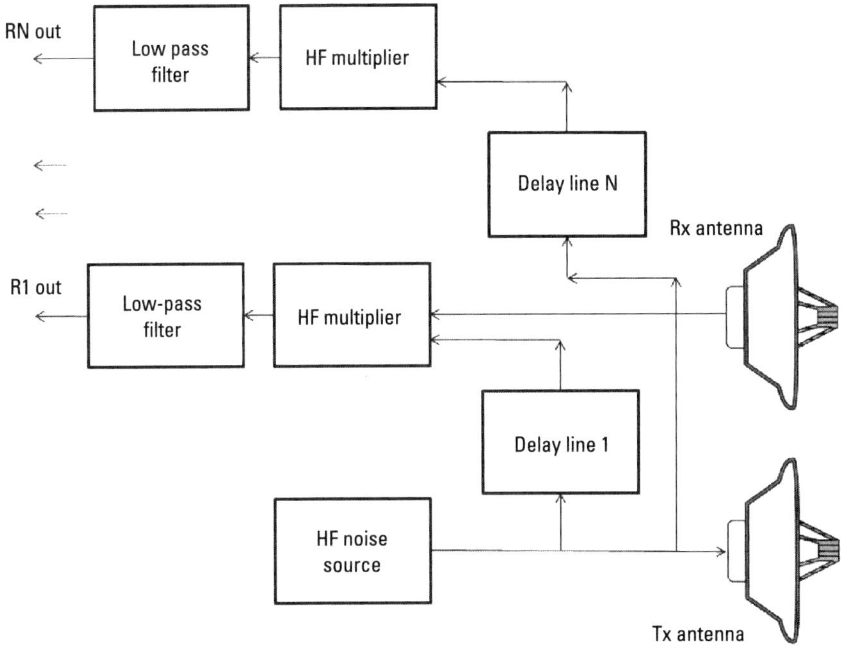

Figure 3.2 Block diagram of the noise radar based on sets of delay lines.

microwave circulator, was amplified by an input amplifier similar to the ones used for signal generation, and then passed via a square low detector (which simply calculated the square of the input voltage) and averaged with time constant *T* similar or equal to the pulse width.

The block diagram of this radar is presented in Figure 3.3. The radar was successfully used for meteorological purposes, detecting areas of clouds and rain, but at that time only amplitude data was used. The radar did not provide information about target velocity as it was very simple.

In the late 1970s more research was done on noise and pseudo noise radars. J. R. Forrest and D. J. Price from University College London worked on a digital correlator for noise radar [7] and later W. Knapp and B. Wigger worked on the use of surface acoustic wave filters (SAW filters) for signal compression in a pseudo noise radar using binary phase shift keying (PNBPSK) [8].

In the last two decades the noise radar concept has been rediscovered and a large number of scientists have begun to work on this kind of radar. The renaissance of noise radar has been caused by the quick development of computing platforms and fast analog/digital converters. Modern high-speed digital signal processors (DSP), programmable logic devices (PLD) equipped with hardware multipliers, and multicore graphical processing units (GPU) make it

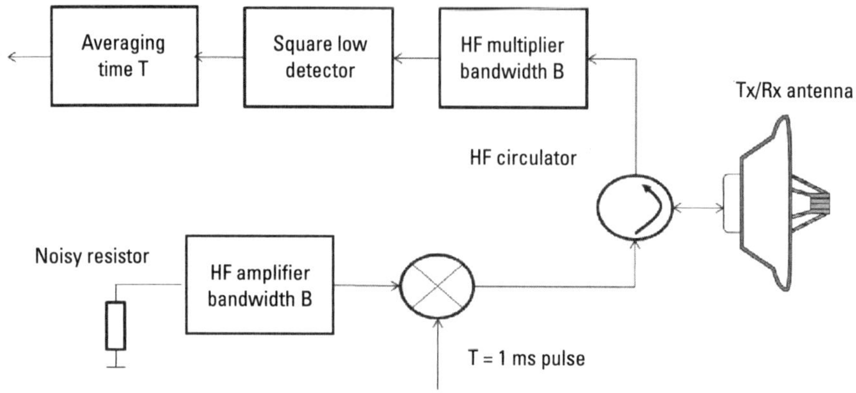

Figure 3.3 Block diagram of a pulse noise radar.

possible to calculate the transmitted and received signal cross-ambiguity function in real time. At present, the noise radar concept is the subject of intensive investigation. Many papers have been published on the short-range surveillance radar [9–11], on the imaging radar working both in SAR [13–28] and ISAR [29] modes, and on ground-penetrating radars [30–32].

Modern noise radars can be divided into several groups depending on the selected criteria. One of the most common criteria is the radar application. According to this criterion, noise radars can be divided into the following main groups: surveillance radars, tracking radars, imaging radars, and ground-penetrating radars, among others. It is also possible to divide the noise radar according to the carrier frequency, bandwidth, and range. The most important criterion in this book is the integration time criterion. According to this particular principle, it is possible to divide noise radar into three major groups [33]: short, medium, and long integration time radars.

3.1 Short-Integration-Time Pulse Noise Radar

A short-integration-time noise radar is a pulse coded radar that, instead of using a classical LFM or phase coded waveform, emits a noise or pseudo-noise waveform signal. Such a radar with incoherent processing has been described in [6]. However, most modern radars rely on coherent processing. This type of radar emits a train of noise-modulated electromagnetic pulses toward the target, applies the pulse compression technique, and detects the target. The range to the target is calculated by estimating the time delay of the return signal, and the radial velocity of the target is estimated using MTD techniques [34]. In terms of properties, the only difference between this type of noise radar and the pulse-coded radar is the level of, and phase instability in, the range sidelobes. If the noise modulation is the same for all pulses within the MTD batch, then the sidelobes are stable and MTD processing will cancel both the ground clutter and its time sidelobes. In the case where a different realization of noise is applied for each pulse, then the main lobe clutter (in the time dimension) will be canceled by MTD processing, while the time sidelobe clutter will remain and decrease the ability to detect small targets in the clutter environment.

For pulse noise radar the integration time is equal to the transmitted pulse length. Due to a short integration time, it is necessary to calculate a single-dimensional correlation function between the transmitted and the received signals. To avoid detection losses caused by Doppler shift, the pulse duration time should be shorter than $\lambda/2v_{max}$. A typical integration time, depending on the radar carrier frequency and the maximum target velocity, is within 1÷100 μs. The advantage of such a radar is that the signal processing is relatively simple, consisting of range compression and MTD processing, and requires relatively low computational power.

3.2 Medium-Integration-Time Continuous Wave Noise Radar

In a medium-integration-time radar target detection is performed using range-Doppler correlation processing. This procedure is very similar to the calculation of the cross-ambiguity function [35–37]. The medium integration time refers to a situation for which the target remains in both the range and Doppler resolution cell during the integration period. The integration time is limited by constraints (2.15) and (2.25), and is usually within the range of 1÷1,000 ms. The maximum integration time depends of course on the radar carrier frequency, the signal bandwidth, the maximum target velocity, and acceleration.

The medium integration time concept may be used in radars equipped with a mechanical or electronically scanned antenna, as well as for those equipped with a multibeam antenna. To avoid strong cross-talk between the transmitter and receiver, the transmitting and receiving antennas have to be separated. A transmitted signal is continuous, at least during integration time. The received signal, which is a delayed, Doppler-shifted copy of the transmitted signal, is divided into segments and processed by a correlation processor. The ranges and radial velocities of targets are evaluated directly by the correlation processor, while the target azimuth is estimated using one of the following techniques: the sigma-delta antenna angle estimation technique [34], the interferometric technique [38, 39], usually called the monopulse estimation [40–44], or the technique of estimating the centroid of power envelope while using a mechanically scanned antenna.

3.3 Long-Integration-Time Continuous Wave Noise Radar

In a long-integration-time radar the target is illuminated for a relatively long time, and in this case the target can migrate between the radar range and range velocity resolution cells. As a rule, the integration time exceeds the constraints (2.15) and (2.25) and is usually in the 1÷10 second range. Such a relatively long integration time may be achieved by the use of a tracking radar or the use of a multibeam radar with a fixed set of antenna beams. In the first approach the receiving antenna has to track the target using either electronic or mechanical beam positioning. The time on target may be almost infinite, but in the case where the tracking process is performed using measured angle data, the integration time is limited by the target position update rate required by the tracker. In the second approach the radar is equipped with an omnidirectional or sector transmitting antenna and a set of directional receiving antennas covering the required surveillance space. Instead of using a multi-antenna receiving system, it is possible to design a multibeam planar or circular receiving antenna using analog or digital beamformers. The signals from targets located in the surveillance space are received simultaneously by a multichannel receiver. The integration time is then limited by the time the target remains within the single antenna beam (2.16), and by the required target update rate (equivalent to the scan time in the classical radar).

Signal processing takes into account the target range and velocity migration, and thus is more elaborate and requires much higher computational power than the signal processing in the medium integration time radar.

3.4 Fundamentals of the Noise Radar Correlation Receiver

As stated in the previous chapter, a noise radar illuminates the target with a noise waveform and listens to the echo. A typical functional diagram of the noise radar is presented in Figure 3.4.

The noise signal is produced in the noise generator. This generator may be constructed using an analog technique, for example, amplifying the thermal noise, or generating the noise digitally by replaying the data samples stored in the memory and converting it to an analog signal using fast digital analog converters (DACs). In general cases, it is assumed that the generator output signal is a complex one. It can be achieved by generating two independent real-valued signals, one for a real part and one for an imaginary part, and treating them as a single complex signal. From a theoretical point of view, the output of the noise generator can be treated as the band-limited noise signal $x_T(T)$, centered around the zero frequency. The bandwidth of the signal B can vary from tens of kilohertz to thousands of megahertz or more. The carrier frequency F is

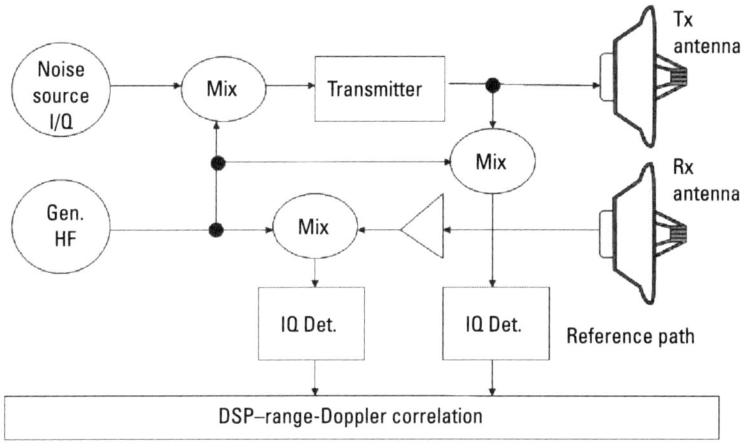

Figure 3.4 An example structure of the coherent noise radar.

generated by a separate, highly stable sine signal generator. The noise signal is upconverted to the carrier frequency in the first mixer. It is worth noting that from a theoretical point of view the mixer performs the operation of multiplying the complex noise signal by the complex harmonic signal. This operation in the frequency domain is equivalent to the simple frequency shift. In this book the technical aspects of upconversion and downconversion are omitted. More information on that subject can be found in [45, 46]. This high-frequency signal generated by the mixer (usually equipped with a bandpass filter) is amplified and emitted using the transmitting antenna. The echo signal arriving to the receiving antenna is also amplified and downconverted to the baseband in the second mixer. It should be pointed out that again the mixing operation can be treated as the frequency shift. At the output of the mixer the complex analog signal is produced. The baseband received signal $x_R(T)$ (consisting of two real-valued signals) is digitized and passed into the correlator. The reference signal for the correlator $x_T(T)$ is produced by downconverting the transmitted signal using a third mixer. In the case where the downconverted transmitted signal is used as a reference signal, both receiving (measurement) and reference signals are passed via an identical (or almost identical) downconverting channel and the amplitude and phase distortion to both signals should be the same. A direct usage of the generated noise signal $x_T(T)$ as the reference signal is also possible, but it would usually lead to large errors in the correlation process, while the upconversion and downconversion processes are a source of amplitude and phase distortion. If we can measure these distortions, it is possible to incorporate a measured frequency response in the processing chain and filter the reference signal using an appropriately designed filter.

In noise radars the target detection process is based on the matched filter concept that is equivalent to the correlation process. The receiver computes the cross-correlation between the transmitted and received signals:

$$y_r(\tau) = \int_{t=0}^{t_i} x_T(t)x_R^*(t-\tau)dt \tag{3.1}$$

where x_T denotes the transmitted signal complex envelope (transmitted signal converted to the baseband), x_R denotes the received signal complex envelope (received signal converted to the baseband), and ti denotes the integration time. While the radar is a device that should estimate the range to the target, (3.1) can be rewritten as:

$$y_r(r) = \int_{t=0}^{t_i} x_T(t)x_R^*\left(t-\frac{2R}{c}\right)dt \tag{3.2}$$

Theoretically the autocorrelation function of the white noise is a Dirac delta function, the infinite peak at zero time delay and zero for other time delays, as presented in Figure 3.5.

Such results are for the infinite bandwidth of the signal and infinite correlation time. In all practical cases the bandwidth of the signal is limited, and correlation time is also limited as stated in (3.2). In such cases the shape of the correlation function is (in approximation) presented in Figure 3.6. The main peak width is approximately $1/T_i$, while the time sidelobe level is approximately $\sqrt{t_iB}$ times below the main peak (in the linear amplitude scale).

A correlation receiver improves the signal-to-noise ratio (S/N) by the factor T_iB, where B is the bandwidth of the transmitted noise signal. This simple correlation processing can be used only in short integration noise radars for

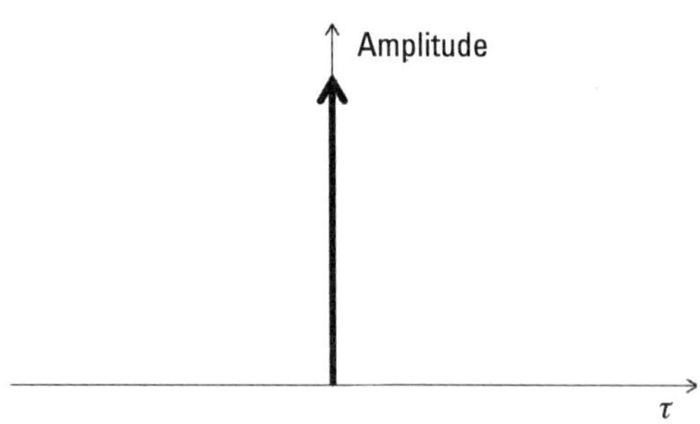

Figure 3.5 Theoretical shape of the noise correlation function.

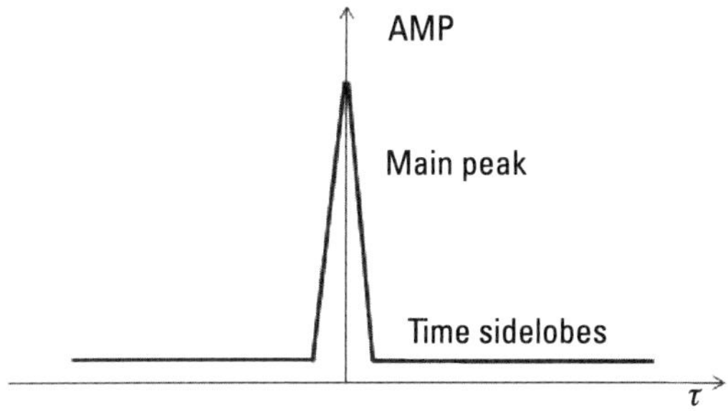

Figure 3.6 Approximation of the correlation function shape of band- and time-limited noise.

slow moving targets, where the Doppler effect during integration time can be negligible. Figure 3.7 presents the example signal produced by the correlator when a T_iB factor is equal to 100, in the case where the transmitted and received signals are identical.

Figure 3.8 shows the output signals produced by a correlator in a situation where only white thermal noise is received. The powers of the received signals are equal in both cases.

The shape of the main peak of the correlation function and the shape of the time sidelobes are related to the shape of the spectrum of the noise signal.

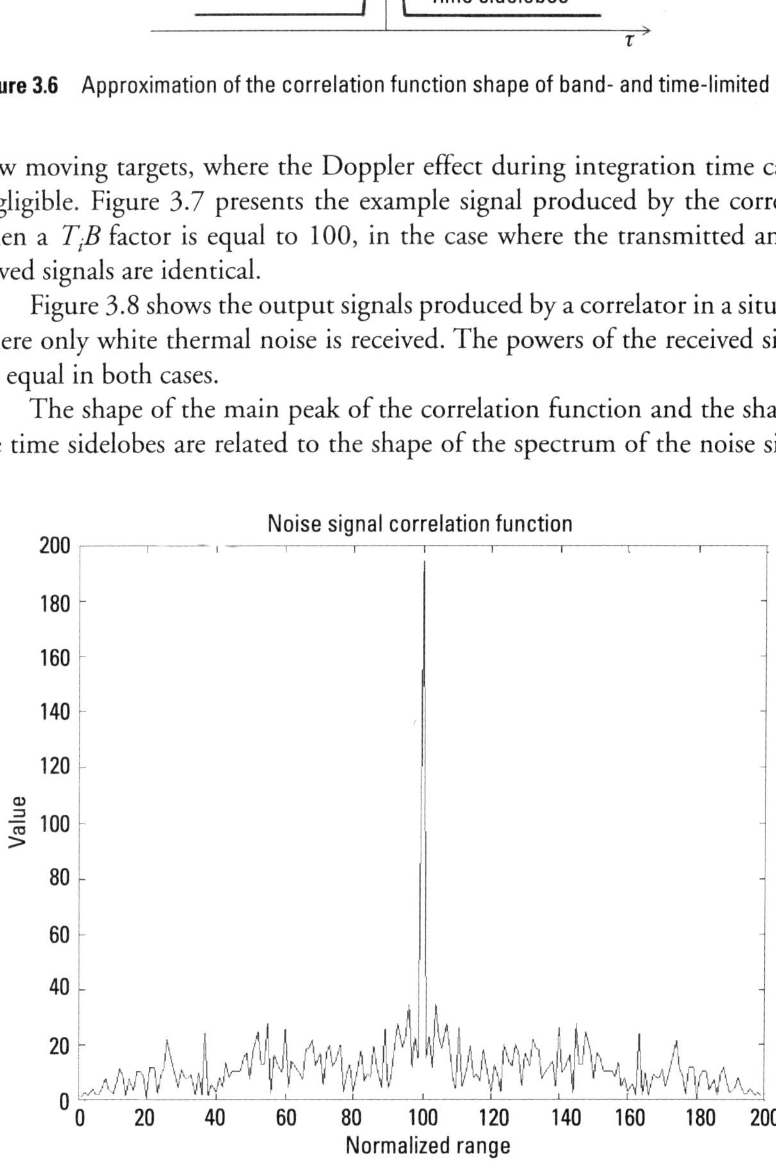

Figure 3.7 Noise signal autocorrelation function.

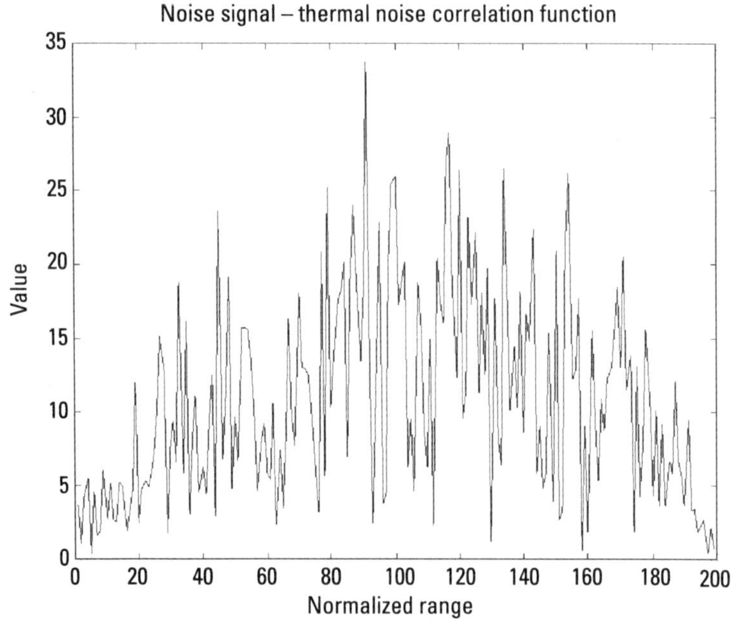

Figure 3.8 Noise signal thermal noise correlation function.

The correlation function (3.2) can also be calculated in the frequency domain using the FFT procedure:

$$y_r(\tau) = FFT^{-1}\left(FFT(x_T(t))FFT^*(x_T(t))\right) \tag{3.3}$$

To analyze the shape of the autocorrelation function of noise, instead of cross-correlation equation (3.3), autocorrelation equation in the frequency form can be used:

$$y_{xx}(\tau) = FFT^{-1}\left(abs(FFT(x_T(t)))^2\right) \tag{3.4}$$

As we can see from (3.4), the autocorrelation function is the inverse Fourier transform of power spectral density of the noise signal. Let us start our consideration from the rectangular spectrum of the noise signal. The inverse Fourier transform of the rectangular signal is in the sinc ($\sin(x)/x$) function, so uses the rectangular signal spectrum (or power density; in this case both signal spectrum and signal power density has a rectangular shape). The sinc function has high sidelobes (up to −13 dB), so such a signal spectrum is not the preferable one for a radar application. Of more use is the triangle power spectrum, for which the autocorrelation function is sinc^2. The first sidelobe is at the level of −26 dB below the main peak, but such a value is still high. The most interesting shape of spectrum is the Gaussian function $\exp(-\alpha x^2)$. In this case the autocorrelation function is also Gaussian and no lobes are present (in theory). In practice the limitation of integration time introduces sidelobes (sometimes referred to as residual fluctuations) as shown in Figures 3.6 and 3.7.

The selected power density function and related autocorrelation functions are summarized in Table 3.1. Please note that in many books [47] it is possible to find more functions that have the Fourier transform in analytic form, but the power density function is a positively defined function (having only nonnegative values) and only such functions can be taken into consideration.

3.5 Target Detection in Noise Radar

In the previous section information on how to process signals using the correlation function was presented. Assuming that a received signal consists only of receiver white noise and echo, the optimal detection (in the mean square sense) is based on correlation processing, but such processing must be followed by an appropriate detector, as presented in Figure 3.9.

To detect the useful signal in most radar systems, the constant false alarm Neyman-Pearson detector [34] is commonly used. This detector compares the signal after correlation processing with the selected threshold D. The hypothesis H_0 (where there is only thermal noise and no useful signal) is assumed when the signal is below the threshold, and hypothesis H_1 (where there is thermal noise plus target echo) is assumed when the signal exceeds the threshold.

When no reflected (echo) signal is received, the signal at the output of a correlation receiver has a Rayleigh probability density function (PDF) due to

Table 3.1
Power Density Functions and Related Autocorrelation Function

Power Density Function	Autocorrelation Function	Comments
$rec\left(\frac{f}{B}\right)$	$B\,sinc(B\tau)$	Rectangular spectrum
$tri\left(\frac{f}{B}\right)$	$B\,sinc^2(B\tau)$	Triangle spectrum
$exp\left(-\pi\frac{f^2}{B^2}\right)$	$\frac{B}{\sqrt{\pi}}exp(-B^2\tau^2)$	Gaussian spectrum
$\frac{B^2}{B^2+(2\pi f)^2}$	$\frac{B}{2}exp(-B\lvert\tau\rvert)$	Lorentzian function of autocorrelation

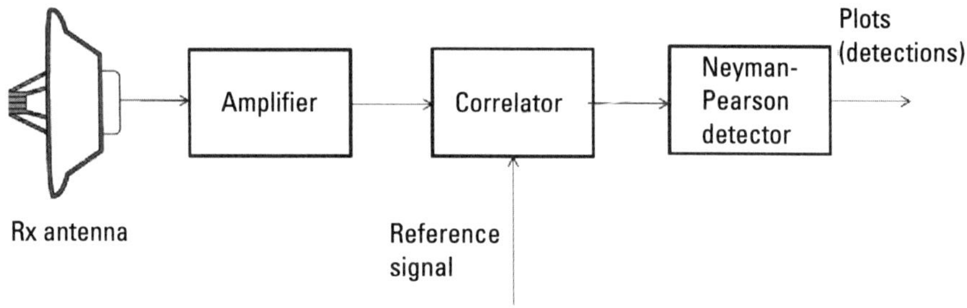

Figure 3.9 The processing chain in noise radar.

the central limit theorem (CLT) and almost regardless of the input noise (or interference) PDF. The CLT is applicable in this case where the discreet implementation of correlation processing is equivalent to the weighted sum of the independent (or almost independent) samples of input noise or interferences.

In Figure 3.10 the threshold value versus the probability of a false alarm (PFA) is plotted. It can be easily seen that for the 10^{-6} probability of a false alarm, the threshold level D_o should be at 12 dB over the correlator output noise level [34]. The probability of detection depends on the mean return signal power and selected threshold level. In Figure 3.11 the detection probability for

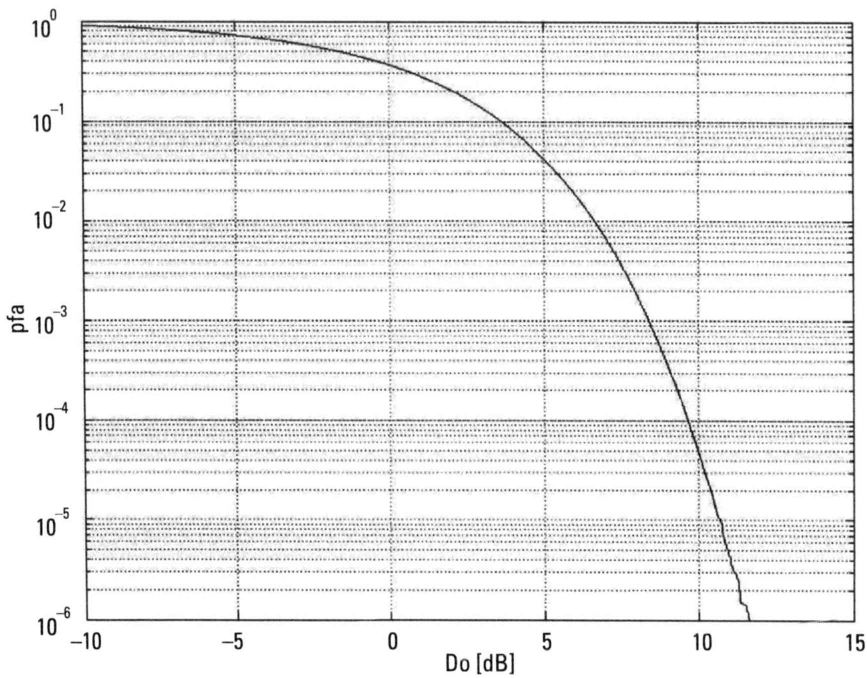

Figure 3.10 The false alarm probability PFA versus threshold level D_0 (Rayleigh distribution).

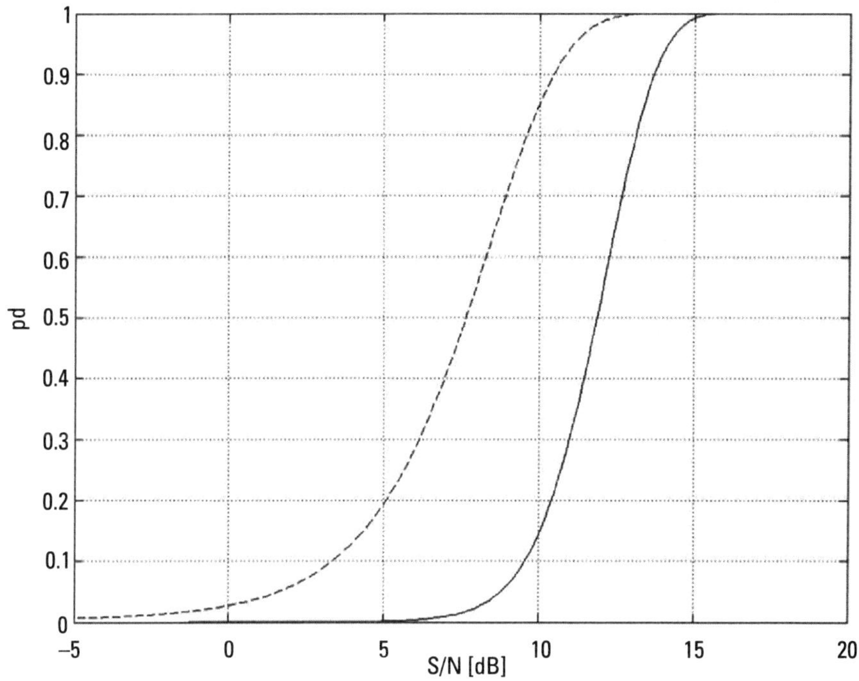

Figure 3.11 The detection probability versus the signal-to-noise ratio (S/N). Solid line: detection threshold 12 dB (PFA ≈ 10–6). Dashed line: detection threshold 8 dB (PFA ≈ 10–3).

a nonfluctuating target (Swerling model case 0 [34]) is plotted. As follows from the figures, the detection probability 0.9 can be reached for a signal to noise level equal to 14 dB (with a threshold chosen for PFA = 10^{-6}).

3.6 Correlation Processing of Moving Target Echoes

The correlation function described by (3.1) can only be used for motionless (stationary) or very slow targets. To detect moving targets, the Doppler frequency shift has to be taken into account in the detection procedure. Let us assume that the transmitted signal can be treated as a narrowband signal. The received signal, reflected from a moving target, is then a delayed and Doppler-shifted copy of the transmitted signal and can be described by (2.19). The complex envelope of the received signal can be expressed by the formula:

$$x_R(t) = \xi_t(t) + Ax_T\left(t - \frac{2R}{c}\right)\exp\left(2\pi j\left(-\frac{2v_oF}{c}\right)t\right) \tag{3.5}$$

where A denotes the complex echo signal amplitude (A = |A| exp(jφR). The optimum detector (in the mean-square sense) is based on the matched filter concept. The output of the filter, matched to the signal echo described by (3.5), can be calculated as an integral:

$$y = \int_{t=o}^{t_i} x_R(t)x_T^*\left(t - \frac{2R}{c}\right)\exp\left(-j2\pi\left(-\frac{2vF}{c}\right)dt\right) \tag{3.6}$$

Of course, the concept of the matched filter can be utilized directly only when the position and velocity of the target are known. To detect a target at an unknown position, it is necessary to construct a bank of filters matched to all possible targets ranges and their velocities. Such an approach leads directly to the range-Doppler correlation function, defined as:

$$y(R,v) = \int_{t=o}^{t_i} x_R x_T^*\left(t - \frac{2R}{c}\right)\exp\left(-j2\pi\left(-\frac{2vF}{c}\right)t\right)dt \tag{3.7}$$

The above relation is very similar to the cross-ambiguity function (2.26), but as opposed to the cross-ambiguity function, the time delay is applied only to the transmitted signal. Such a transformation form is more convenient for practical implementation in digital techniques, and will be further referred to as a range-Doppler cross correlation function.

The relation (3.7) can be treated as a set of one-dimensional correlation functions of the received signal and the Doppler-shifted transmitted signal, or as a set of Fourier transforms of the multiplication of the received signal and a complex conjugation of the time-shifted transmitted signal.

Figure 3.12 presents an example diagram of the absolute value of the range-Doppler cross-correlation function between the received and transmitted signals for a single target at the distance of 10 km and radial velocity of 100 m/s, computed directly using (3.7). It is easy to notice the presence of relatively high-frequency sidelobes, caused by a boxcar window following integration limits. To decrease the sidelobe level, other time windows can be used. The windowing can be applied either at the transmission side (by changing the amplitude of the transmitted signal) or during signal processing. The second approach leads to the concept of nonmatched filtering, described by the following formula:

$$y(R,v) = \int_{t=o}^{t_i} w(t)x_R x_T^*\left(t - \frac{2R}{c}\right)\exp\left(-j2\pi\left(-\frac{2vF}{c}\right)t\right)dt \tag{3.8}$$

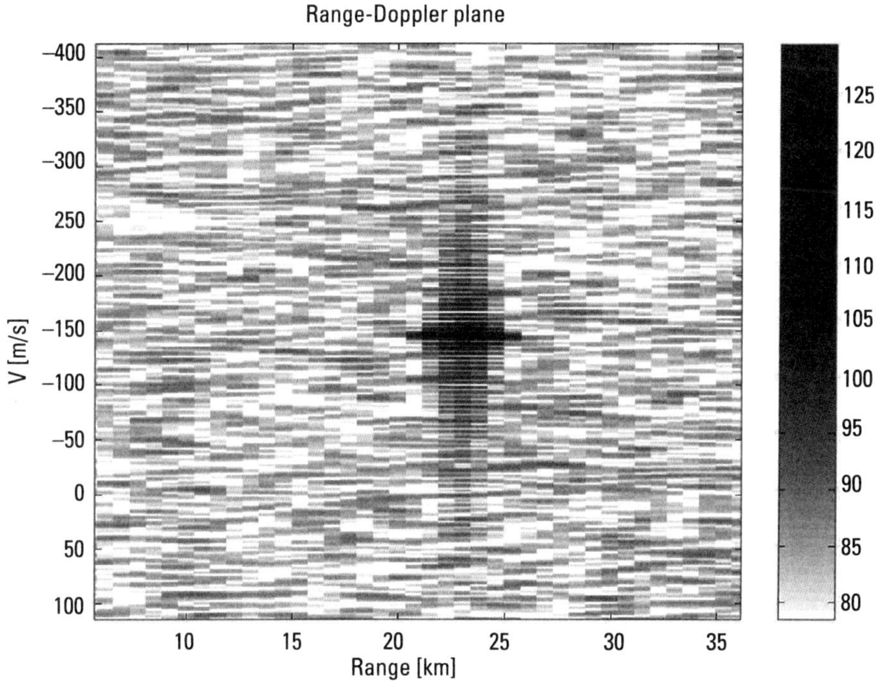

Figure 3.12 An example of the range-Doppler cross-correlation function of noise radar return: no time windowing.

where $w(t)$ is a windowing function. An example result of the application of the Hamming window to range-Doppler correlation processing is presented in Figure 3.13. The first sidelobe level is decreased from –13 dB to -60 dB.

3.7 The Range Equation Formulated for Noise Radar

Let us consider a noise radar equipped with a transmitting antenna with gain G_T and a receiving antenna with gain G_R. This radar transmits the electromagnetic signal of power P_T and illuminates a target characterized by cross-section So located at a distance R from the radar. The received echo signal power is given by the expression:

$$P_R = \frac{P_T G_T G_R S_o \lambda^2}{(4\pi)^3 L R^4} \tag{3.9}$$

To detect the reflected signal in the presence of the thermal noise, the correlation process (matched filtering described by (3.8)) is used. In this case the

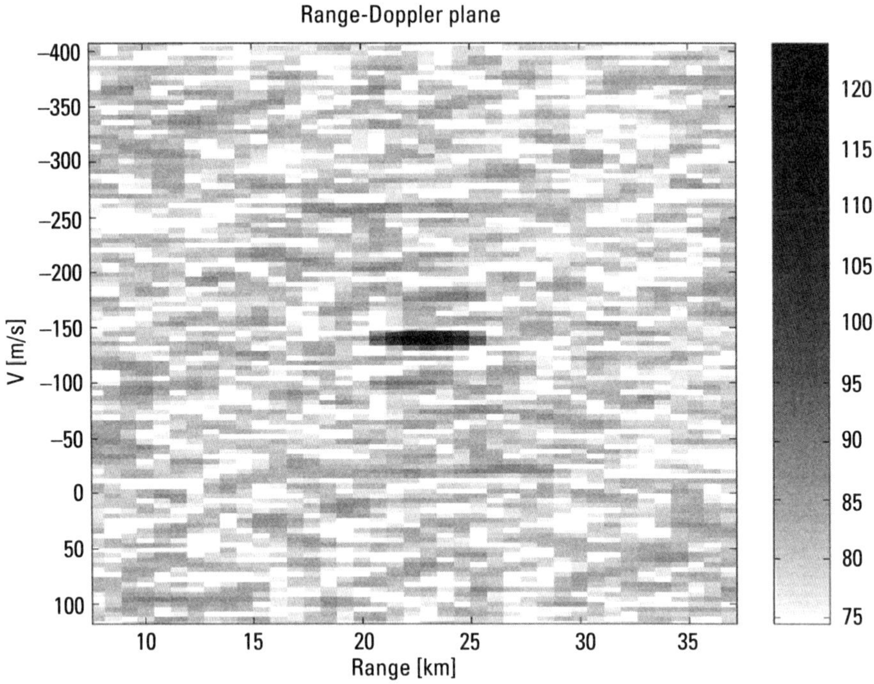

Figure 3.13 An example of the range-Doppler cross-correlation function of noise radar return: Hamming window.

received signal can be detected when its power is higher than the power of the thermal noise P_N multiplied by a detectability factor. As a result, the target will be detected by a noise radar if the following constraint is fulfilled:

$$P_R \geq P_N \cdot D = kT_R BD \tag{3.10}$$

From the theory presented earlier, it follows that the value of the detectability factor is equal to the ratio of the threshold chosen for the desired probability of the false alarm rate (see Figure 3.10) and the correlation gain equal to the integration time and bandwidth product as stated here:

$$D = \frac{D_0}{t_i B} \tag{3.11}$$

The maximum value of the detection range for the noise radar is:

$$R_{\max} = \sqrt[4]{\frac{P_T G_T G_R S_0 \lambda^2 t_i}{(4\pi)^3 L D_0 k T_R}} \tag{3.12}$$

The detection range can be increased by increasing the transmitted power P_T, the antenna gain G, and the integration time T_i. For a medium integration time the time-bandwidth product T_iB is limited by the range migration effect. This means that the following none quality must be fulfilled:

$$t_i B \leq \frac{c}{2V_{\max}} \tag{3.13}$$

Assuming the maximum target velocity at the level of Mach 3 (1,000 m/s), the maximum value of the time-bandwidth product is limited to 150,000. The maximum processing gain is then equal to 51.7 dB. The use of a windowing function will decrease this value by a few decibels. If the noise radar is used for detecting or imaging slower targets, then it is possible to increase the correlation gain, as stated in Table 3.2.

3.8 Long-Correlation-Time Noise Radar Processing

The integration (correlation) time t_i is one of the most important parameters that can influence the sensitivity of a noise radar. The integration time in the medium-integration-time radars, where signal processing is based on (3.7), is limited by the range and velocity cell migration effects, described by constraints (2.15) and (2.25) to the values presented in Table 3.2. To achieve a longer correlation time, it is necessary to incorporate range and Doppler cell migration effects into the detection algorithm. Let us consider at the beginning the range cell migration problem. Assuming a constant target radial velocity v, the range to the target can be written as $R = R_o + vt$. In (3.5) the influence of the target velocity is limited only to the Doppler effect. As stated in (2.19), except for the

Table 3.2
Maximum Integration Gain Versus Maximum Target Velocity

Maximum Target Velocity [m/s]	Maximum Integration Gain (Including Windowing) [dB]	Comments
1	80	Walking human
10	70	Bike
50	63	Car
300	55	Passenger plane
500	53	Jet fighter
1,000	50	Fast jet plane/rocket
2,000	47	Fast rocket

Doppler shift, the received signal is a delayed copy of a time-scaled transmitted signal. The matched filter equation incorporating the time-scaling of the received signal is now in the form:

$$y(R,v)=\int_0^{t_i} x_R(t)\cdot x_T^*\left(t-\frac{2R+2vt}{c}\right)\cdot\exp\left(4\pi jvt\,F/c\right)dt \tag{3.14}$$

This equation can be rewritten in the following form:

$$y(R,v)=\int_0^{t_i} x_R(t)\cdot x_T^*\left(t\left(1-\frac{2v}{c}\right)-\frac{2R}{c}\right)\cdot\exp\left(4\pi jvt\,F/c\right)dt \tag{3.15}$$

It is now clearly seen that the reference signal is now stretched according to the ratio of target and light velocity. The computational complexity involved in calculating a range-Doppler cross correlation function is now much higher than the computational cost of calculating the final result using (3.8), while it is necessary to time-scale the transmitted signal. Time scaling, also referred to as stretch processing, may be performed in the time domain by resampling a transmitted signal or in the frequency domain by using the chirp transform [47, 48].

3.9 Stretch Processing

To obtain long time integration, it is necessary to perform the stretch processing/time-scale change of the reference signal according to (3.12). The idea of applying stretch processing as time transformation is not new. Time scaling, time slowdown or speed-up, and even time reverse techniques were introduced in the second half of the twentieth century [49]. For time scaling the analog dispersive delay line was used and time scaling was performed that time for short pulses only.

In a modern noise radar correlation processing is performed digitally, so the time scaling can be achieved by resampling the reference signal. The resulting reshaping of the reference signal spectrum should not introduce phase and frequency errors. The quality of stretch processing is fundamental for obtaining a high integration gain and long detection range. The above conclusion is also valid for other types of radars. In pulse radar simplified range migration processing can be performed after pulse compression by the coherent integration of the signal over several range cells [50]. Stretch processing in the FMCW radar has been considered in [51]. In this case the stretch processing may be implemented effectively with the use of a group of subband filters (frequency-dependent delay lines). The applications of stretch processing have also been presented in [52–54].

In noise radars stretch processing can be implemented in many different ways. Among others, the three most popular algorithms that are frequently used are: linear interpolation between samples, cubic spline interpolation, and resampling based on the pairing of chirp transform and inverse FFT. The results obtained by using the algorithms mentioned above are close to each other.

It should be pointed out that stretch processing should be performed independently for each velocity resolution cell. However, such an approach requires very high computational power. To implement a long integration time concept in real-time applications, it is essential to make some simplifications. In most cases it is sufficient to stretch the reference signal not for each Doppler resolution cell but for the group of velocity cells occupying the velocity strip width Δv. The velocity strip width Δv should satisfy the constraint (2.15). In other words, the strip width Δv should be smaller than $c/2BT_i$. Thus, in order to calculate the whole range-Doppler correlation plane it is necessary to perform $N = 4BT_i v_{max}/c$ stretching operations and calculate the range-Doppler correlation in each narrow velocity interval.

The effectiveness of stretch processing verified using a Monte Carlo simulation is presented in Figures 3.14 through 3.17. Three moving point objects were used in simulations. The first object had a range velocity of –270 m/s and a range of 5.8 km, the second object had a velocity of 6 m/s and a range of 13.4 km, and the third object had a velocity of 705 m/s and a range of 25 km. All echoes had the same strength. The results of the long-time range-Doppler processing of the 1-second data record (100-kHz bandwidth) are presented in Figure 3.14. Without stretch processing only the second target echo is properly focused, as it is almost stationary. The first and the third target echoes are defocused in both range and velocity. The integration gain of the first echo is 3 dB smaller than the integration gain of the almost stationary target. For the third target, the integration loss is greater than 10 dB.

The range cross-section of all plots is presented in Figure 3.15. It is easy to see that defocusing is strictly related to signal losses. Additionally, high-velocity targets, for which the range migration effect occurs, are smeared not only in range but also in velocity. This effect is due to the fact that the object remained in the single resolution cell for much shorter than the integration time T_i, and shortening the effective integration time (limited in this case by range walk) decreased the velocity resolution.

The concept of dividing the full velocity range into several intervals requires the calculation of several range-Doppler cross-correlation functions, each using a reference signal stretched according to velocity $nc/2BT_i$, where n is the index of the segment $-N/2 < n < N/2$).

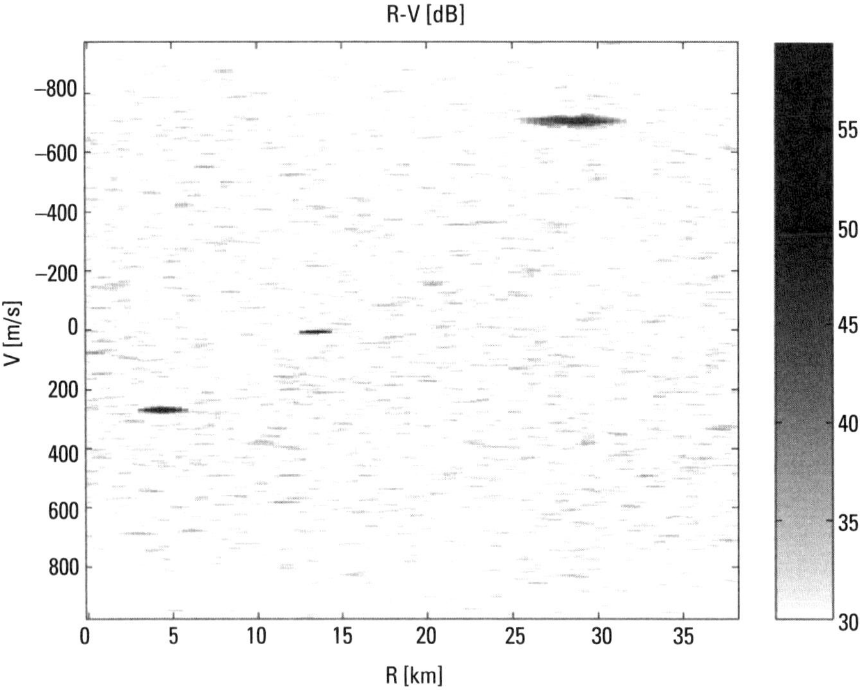

Figure 3.14 The absolute value of the range-Doppler cross-correlation function without stretch processing, three echoes (grayscale in decibels).

Figure 3.16 presents the absolute value of the whole range-Doppler cross-correlation function calculated using a reference signal stretched according to the velocity of the third echo. The cubic spline algorithm has been used in this example. It is obvious that the third target echo is focused, while the other echoes are unfocused and attenuated.

The final results of the segment-based stretch processing are presented in Figure 3.17. The correlation plane has been divided into N velocity segments, and each of them has been stretch-processed with the segment middle velocity as a reference. All three echoes are focused and the amplitudes are now almost equal.

Consequently, for extended integration time the assumption of constant target radial velocity may not be valid. The target radial velocity changes (range acceleration) can cause both velocity cell migration and additional range cell migration. If integration time exceeds the time limit described by (2.25) (velocity cell migration), or exceeds the limit described by the following formula:

$$t_i < \sqrt{\frac{2\Delta r}{a_{\max}}} = \sqrt{\frac{c}{Ba_{\max}}} \tag{3.16}$$

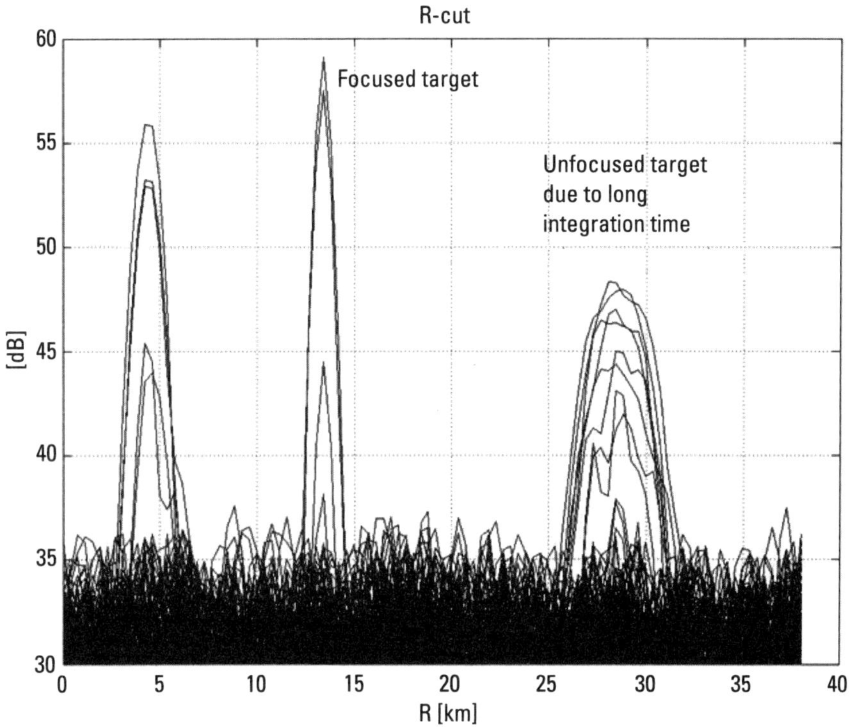

Figure 3.15 Range cross-section of plots, amplitude in the decibel scale.

then the target echo will be defocused. Those formulas describe the limits after which a more complex approach is needed, adding an extra dimension of acceleration.

3.10 Processing with Radial Acceleration

The extension of the integration time requires the introduction of radial acceleration into the target motion model. For the constant radial acceleration the range to the target can be expressed as $R = R_o + vt + \frac{at^2}{2}$. The matched filter concept leads now to a three-dimensional range-Doppler-acceleration cross-correlation function defined by:

$$y(R,v,a) = \int_0^{t_i} x_R(t) \cdot x_T^*\left(t - \frac{2R + 2vt + at^2}{c}\right) \cdot \exp\left(j2\pi(2vt + at^2)F/c\right)dt \qquad (3.17)$$

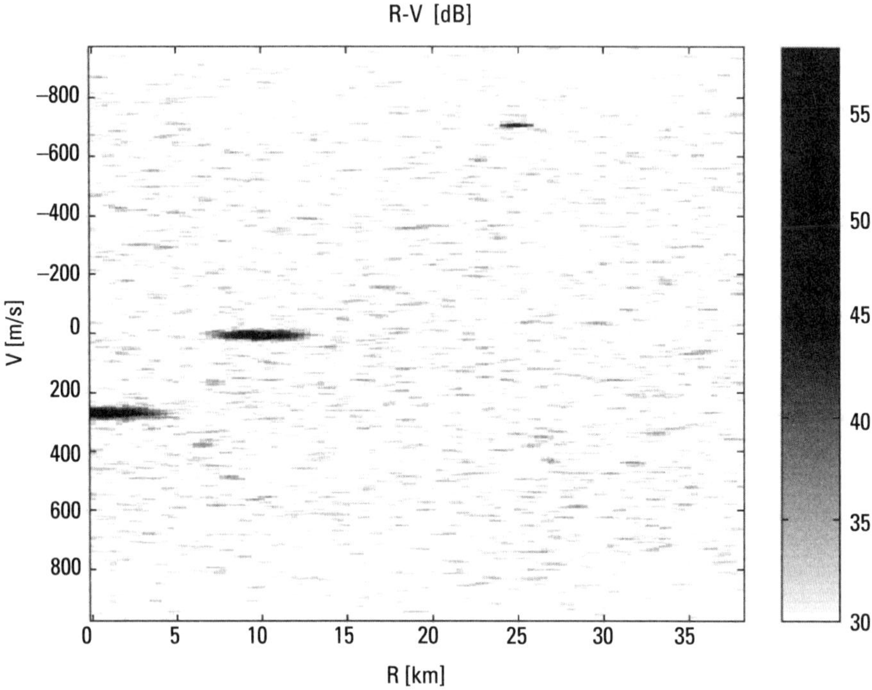

Figure 3.16 Absolute value of the range-Doppler cross-correlation function; reference signal stretched according to the third target echo (grayscale in decibels).

Similar to the previous case, effective processing can be performed by dividing the velocity-acceleration space into the rectangular segments with a velocity interval not greater than $\Delta v < c/2BT_i$ and an acceleration interval $\Delta a < 2\lambda / t_i^2$. For each segment the cross-correlation function can be computed using conventional range-Doppler processing (3.8) with the modified reference function:

$$x_T^m(t, v_s, a_s) = x_T\left(t - \frac{2v_s t + a_s t^2}{c}\right) \cdot \exp\left(-j2\pi(2v_s t + a_s t^2)F/c\right) \quad (3.18)$$

The limitation of the integration time in long-correlation-time noise radars is caused now by the rate of acceleration changes, the time that the target remains in the antenna beam, and the required target refresh ratio. In typical radar systems this time limit takes values from the range of 1÷10 seconds. It is worth noting that for a long correlation time it is possible to estimate not only the target position and range velocity, but also the target range acceleration, so the target update rate can be relatively low without the risk of the radar tracker losing track or making an error in the target-plot association procedure.

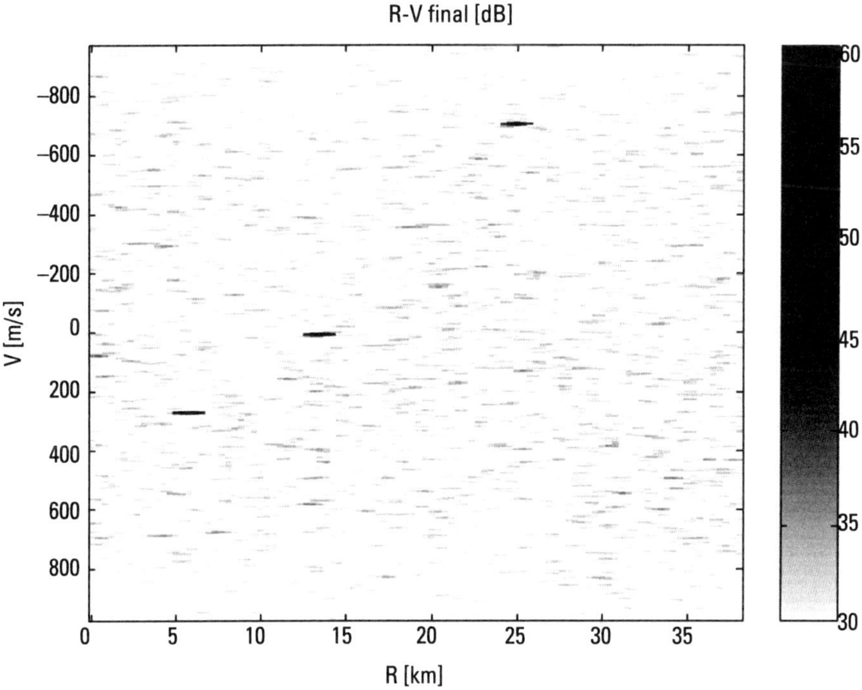

Figure 3.17 Range-Doppler correlation plane with stretch processing.

The idea of acceleration estimation in a radar technique is not new [59]. The accuracy of the acceleration estimation depends on the target observation time. In classical air traffic control radars such time is very short, so the accuracy of the estimation of both velocity and acceleration is very low. In FMCW radars the observation time is usually much longer, so more effort was assigned to this problem [55, 57, 60].

The estimation of target acceleration can be performed both in long and medium correlation time noise radars [56]. The coordinates of the maximum of the range-Doppler correlation function are the estimates of the target coordinates (range R_T and velocity v_T). The target acceleration can be estimated by finding the coordinate of the maximum of the single-dimensional acceleration transform [56] calculated for the estimated target range and velocity:

$$y_a(a) = \int_{-t_i/2}^{t_i/2} x_R(t) \cdot x_T^*\left(t - 2\frac{R_t}{c}\right) \cdot \exp\left(4j\frac{v_t t + at^2/2}{\lambda}\right) dt \tag{3.19}$$

The example results of the estimation of the target acceleration are presented in Figures 3.18 through 3.20. A single moving target placed at a distance of 15.5 km with a range velocity of 430 m/s and a range acceleration of 8.8 m/

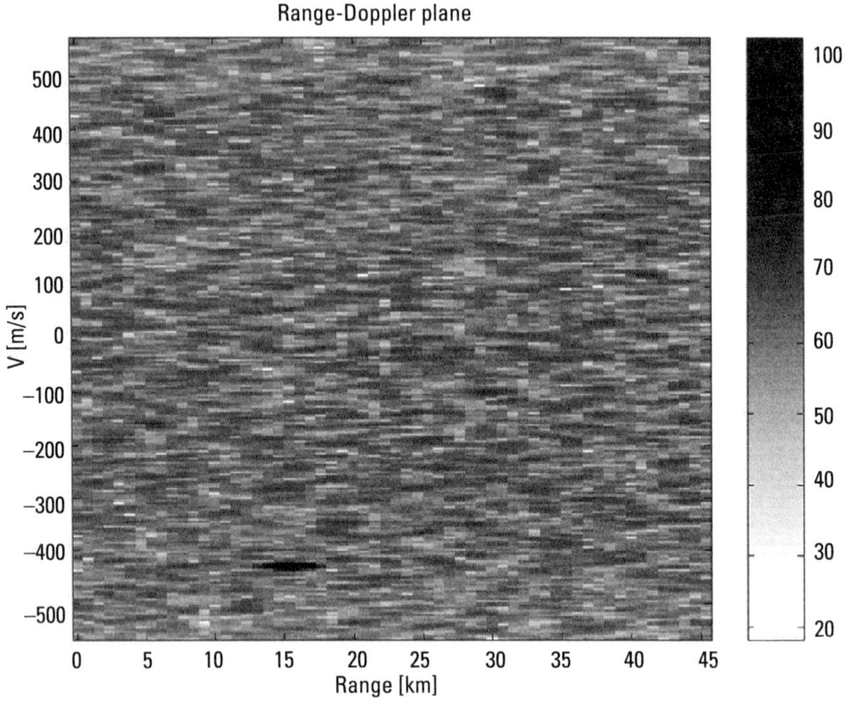

Figure 3.18 The values of the absolute value of the range-Doppler cross-correlation function for an accelerating target (grayscale in decibels).

s^2 was used in the simulation. In Figure 3.18 the results of the range-Doppler processing (2-D processing) are presented. The echo is well focused in both range and velocity, while the acceleration is small enough and fulfills the constraint (2.25).

The absolute values of a velocity-acceleration cross correlation function (3.19) for the estimated target range are presented in Figure 3.19, and the results of the acceleration processing described by (3.18) are presented in Figure 3.20. The estimate acceleration (the coordinates of the maximum acceleration transform) agreed very well with the model.

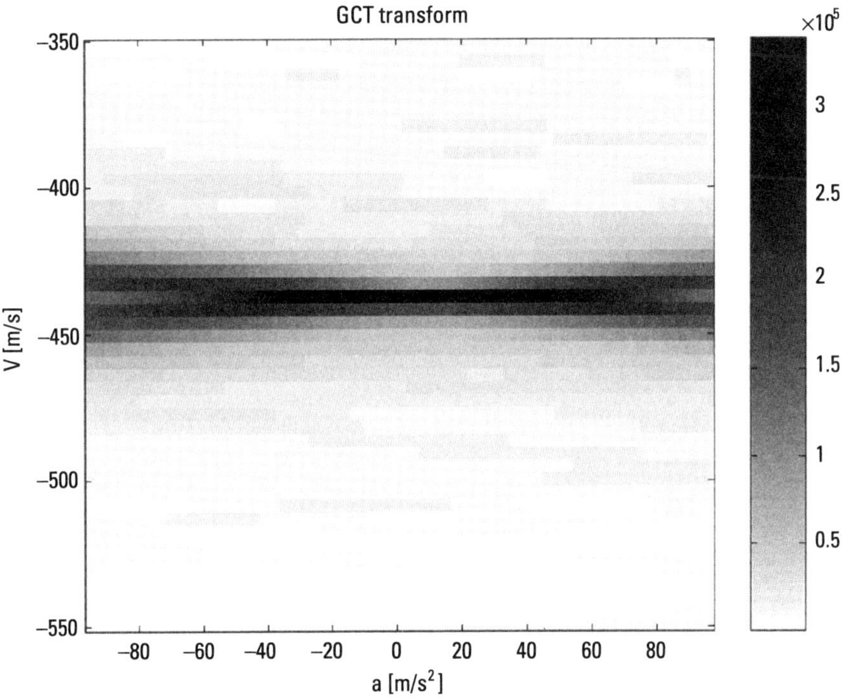

Figure 3.19 The target velocity-acceleration plane.

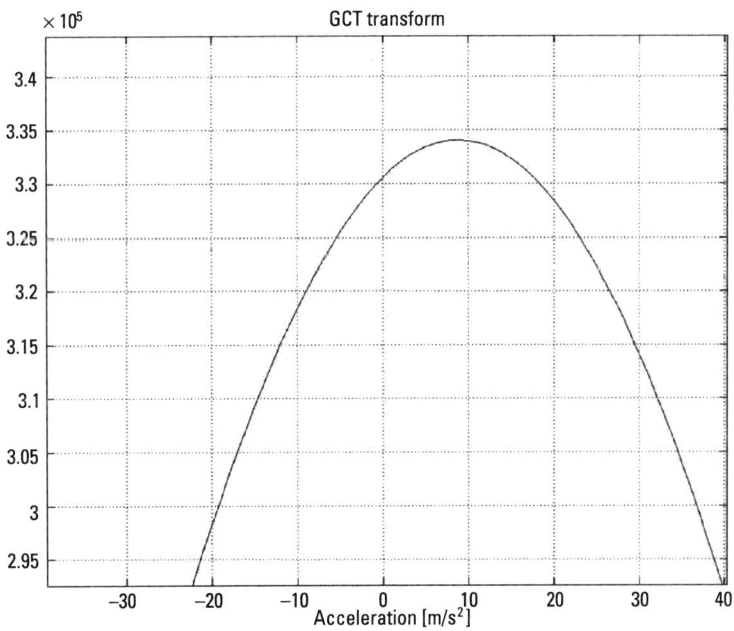

Figure 3.20 The result of acceleration transform described by (3.19).

References

[1] Guosui, L., G. Hong, and S. Weimin, "Development of Random Signal Radars," *IEEE Transactions on Aerospace and Electronic Systems*, Vol. 35, No. 3, July 1999, pp. 770–777.

[2] Horton, B. M., "Noise-Modulated Distance Measuring System," *Proc. IRE*, V0147, May 1959, pp. 821–828. Reprinted in *Radars, Vol. 7, CW and Doppler Radar*, D. K. Barton, (ed.), Dedham, MA: Artech House, 1978.

[3] Cooper, G. R., and C. D. McGillem, "Random Signal Radar," School Electr. Eng., Purdue Univ., Final Report, TREE67-11, June 1967.

[4] Craig, S. E., W. Fishbein, and O. E. Rittenbach, "Continuous-Wave Radar with High Range Resolution and Unambiguous Velocity Determination," *IRE Trans. Mil. ElecTronics*, Vol. MIL 6, No. 2, April 1962, pp. 153–161. Reprinted in *Radars, Vol. 7, CW and Doppler Radar*, D. K. Barton, (ed.), Dedham, MA: Artech House, 1978.

[5] Grant, M. P., G. R. Cooper, and A. K. Kamal, "A Class of Noise Radar Systems," *Proceedings of The IEEE*, Vol. 51, No. 7, July 1963, pp. 1060–1061.

[6] Krehbiel, P. R., and M. Brook, "A Broad-Band Noise Technique for Fast-Scanning Radar Observations of Clouds and Clutter Targets," *IEEE Transactions on Geoscience Electronics*, Vol. 17, No. 4, October 1979, pp. 196–204.

[7] Forrest, J. R., and D. J. Price, "Digital Correlation for Noise Radar Systems," *ElecTronics Letters*, Vol. 14, No. 18, August 31, 1978, pp. 581–582.

[8] Knapp, W., and B. Wigger, "A Pseudo Noise Modulated CW-Radar with Saw-Convolver for Signal Processing," *17th European Microwave Conference*, September 7–11, 1987, pp. 217–221.

[9] Malanowski, M., and K. Kulpa, "Detection of Moving Targets with Continuous-Wave Noise Radar: Theory and Measurements," *IEEE Transactions on Geoscience and Remote Sensing*, Vol. 50, No. 9, September 2012, pp. 3502–3509.

[10] Malanowski, M., and P. Roszkowski, "Bistatic Noise Radar Using Locally Generated Reference Signal," *2011 Proceedings InTernaTional Radar Symposium (IRS)*, September 7–9, 2011, pp. 544–549.

[11] Kalinin, V. I., and V. V. Chapursky, "Wideband Noise Radar for Detection of Slow Moving Objects," *2004 14Th InTernational Crimean Conference on Microwave and Telecommunication Technology (CriMico 2004)*, September 13–17, 2004, pp. 695–696.

[12] Thayaparan, T., M. Dakovic, and L. Stankovic, "Mutual Interference and Low Probability of Interception Capabilities of Noise Radar," *IET Radar, Sonar & Navigation*, Vol. 2, No. 4, August 2008, pp. 294–305.

[13] Garmatyuk, D., and R. M. Narayanan, "Ultrawide-Band Noise Synthetic Radar: Theory and Experiment," *IEEE Antennas Propagat. Soc. Int. Symp. 1999*, Vol. 3, Orlando, FL, July 1999, pp. 1764–1767.

[14] Davis, M. E., "Technical Challenges in Ultra Wide-Band Radar Development for Target Detection and Terrain Mapping," *Proc. IEEE 1999 Radar Conf.*, Boston, MA, April 1999, pp. 1–6.

[15] Maslikowski, L., and K. Kulpa, "Bistatic Quasi-Passive Noise SAR Experiment," *2010 11th International Radar Symposium (IRS)*, June 16–18, 2010, pp. 1–3.

[16] Maslikowski, L., et al., "Preliminary Results of Ground-Based Noise SAR Experiments," 2010 8th European Conference on Synthetic Aperture Radar (EUSAR), June 7–10, 2010, pp. 1–4.

[17] Maslikowski, L., M. Malanowski, and K. Kulpa, "Stepped Frequency Ground-Based Noise SAR Demonstrator," *2011 3rd International Asia-Pacific Conference on Synthetic AperTure Radar (APSAR)*, September 26–30, 2011, pp. 1–2.

[18] Xu, X., and R. M. Narayanan, "FOPEN SAR Imaging Using UWB Step-Frequency and Random Noise Waveforms," *IEEE Transactions on Aerospace and ElecTronic Systems*, Vol. 37, No. 4, October 2001, pp. 1287–1300.

[19] Lukin, K., P. Vyplavin, and N. Kolchigin, "Noise Waveform SAR Imaging in Antenna Near Zone," *International Radar Symposium 2006*, Krakow, Poland, May 24–26, 2006, pp. 303–306.

[20] Tarchi, D., et al., "Short Range Imaging Applications of Noise Radar Technology," *Proc. of EUSAR 2000*, Munich, Germany, May 23–24, 2000, pp. 361–364.

[21] Garmatyuk, D. S., and R. M. Narayanan, "SAR Imaging Using a Coherent Ultra Wideband Random Noise Radar," in *Radar Processing, Technology, and Applications IV*, W. I. Miceli, (ed.), *Proceedings of SPIE*, Vol. 3810, Denver, CO, July 1999, pp. 223–230.

[22] Soumekh, M., "Reconnaissance with Ultra Wideband UHF Synthetic Aperture Radar," *IEEE Signal Proc. Mag.*, Vol. 12, No. 4, July 1995, pp. 21–40.

[23] Lukin, K. A., "Ground Based Noise-Waveform-SAR for Monitoring of Chernobyl Sarcophagus," *Proc. of IRS 2005*, Berlin, Germany, September 6–8, 2005, pp. 655–659.

[24] Garmatyuk, D. S., and R. M. Narayanan, "SAR Imaging Using Fully Random Bandlimited Signals," *IEEE Antennas and Propagation Society International Symposium*, Vol. 4, 2000, pp. 1948–1951.

[25] Mogila, A. A., et al., "Ka-Band Noise SAR Simulation," *4th International Kharkov Symposium on Physics and Engineering of Millimeter and Sub-Millimeter Waves*, Vol. 1, June 4–9, 2001, pp. 441–443.

[26] Duan, Y., Z. Yu, and Y. Zhang, "Research Progress of Noise Radar Technologies," *2011 3rd International Asia-Pacific Conference on Synthetic Aperture Radar (APSAR)*, September 26–30, 2011, pp. 1–4.

[27] Tarchi, D., et al., "SAR Imaging with Noise Radar," *IEEE Transactions on Aerospace and Electronic Systems*, Vol. 46, No. 3, July 2010, pp. 1214–1225.

[28] Gu, X., Y. Zhang, and X. Zhang, "Stepped Frequency Random Noise UWB Radar Signal," *2011 3rd International Asia-Pacific Conference on Synthetic Aperture Radar (APSAR)*, September 26–30, 2011, pp. 1–4.

[29] Bell, D. C., and R. M. Narayanan, "ISAR Turntable Experiments Using a Coherent Ultrawide Band Random Noise Radar," *IEEE Antennas Propagat. Soc. Int. Symp.*, Orlando, FL, July 1999, pp. 1768–1771.

[30] Xu, Y., et al., "Polarimetric Processing of Coherent Random Noise Radar Data for Buried Object Detection," *IEEE Transactions on Geoscience and Remote Sensing,* Vol. 39, No. 3, March 2001, pp. 467–478.

[31] Reeves, B. A., and W. B. Muller, "Traffic-Speed 3-D Noise Modulated Ground Penetrating Radar (NM-GPR)," *2012 14th International Conference on Ground Penetrating Radar (GPR),* June 4–8, 2012, pp. 165–171.

[32] Muller, W. B., and B. A. Reeves, "Comparing Traffic Speed Deflectometer and Noise-Modulated Ground Penetrating Radar Data for Rapid Road Pavement Investigations," *2012 14Th International Conference on Ground Penetrating Radar (GPR),* June 4–8, 2012, pp. 502–509.

[33] Kulpa, K. S., Z. Czekala, and M. Smolarczyk, "Long-Integration-Time Noise Surveillance Radar," *First International Workshop on The Noise Radar Technology NRTW-2002,* September 18–20, 2002, pp. 238–243.

[34] Skolnik, M. I., *Radar Handbook,* 3rd ed., New York: McGraw-Hill, 2008.

[35] Rendas, M. J. D., and J. M. F. Moura, "Ambiguity in Radar and Sonar," *IEEE Transactions on Signal Processing,* Vol. 46, No. 2, February 1998, pp. 294–305.

[36] Griffiths, H. D., et al., "Measurement and Analysis of Ambiguity Functions of Off-Air Signals for Passive Coherent Location," *Electronics Letters,* Vol. 39, No. 13, June 26, 2003, pp. 1005–1007.

[37] Auslander, L., and R. Tolimieri, "Characterizing the Radar Ambiguity Functions," *IEEE TransacTions on InformaTion Theory,* Vol. 30, No. 6, November 1984, pp. 832–836.

[38] Thomson, A. R., J. M. Moran, and J. G. W. Swenson, *Interferometry and Synthesis in Radio AsTronomy,* New York: John Wiley & Sons, 1986.

[39] Wohlleben, R., H. Mattes, and T. Krichbaum, *Interferometry in Radioastronomy and Radar Techniques,* New York: Springer, 2001.

[40] Nielsen, R. O., "Accuracy of Angle Estimation with Monopulse Processing Using Two Beams," *IEEE Transactions on Aerospace and Electronic Systems,* Vol. 37, No. 4, October 2001, pp. 1419–1423.

[41] Rytel-Andrianik, R., and K. Kulpa, "Monopulse Angle Estimation with Three Sub-Apertures," *2011 Proceedings International Radar Symposium (IRS),* September 7–9, 2011, pp. 218–223.

[42] Moon, S. -H., et al., "Monopulse Angle Estimation with Constrained Adaptive Beamforming Using Simple Mainlobe Maintenance Technique," *2003 IEEE Military Communications Conference (MILCOM'03),* Vol. 2, October 13–16, 2003, pp. 1365–1369.

[43] Galy, J., E. Chaumette, and P. Larzabal, "Joint Detection Estimation Problem of Monopulse Angle Measurement," *IEEE Transactions on Aerospace and Electronic Systems,* Vol. 46, No. 1, January 2010, pp. 397–413.

[44] Wang, Z., et al., "Angle Estimation for Two Unresolved Targets with Monopulse Radar," *IEEE Transactions on Aerospace and Electronic Systems,* Vol. 40, No. 3, July 2004, pp. 998–1019.

[45] Tsui, J., *Digital Techniques for Wideband Receivers*, Norwood, MA: Artech House, 2001.

[46] Pace, P. E., *Advanced Techniques for Digital Receivers*, Norwood, MA: Artech House, 2000.

[47] Oppenheim, A. V., and R. W. Schafer, *Digital Signal Processing*, Englewood Cliffs, NJ: Prentice Hall, 1975.

[48] Cetin, M., and A. D. Lanterman, "Region-Enhanced Passive Radar Imaging," *Proc. IEE Radar, Sonar Navigation*, Vol. 152, No. 3, June 3, 2005 pp. 185–194.

[49] Caputi, W. J., "Stretch: A Time-Transformation Technique," *IEEE Transactions on Aerospace and Electronic Systems*, Vol. AES-7, No. 2, March 1971, pp. 269–278.

[50] Jun, W., Z. Shouhong, and B. Zheng, "On Motion Compensation for Weak Radar Reflected Signal Detection," *6th International Conference on Signal Processing*, Vol. 2, August 26–30, 2002, pp. 1445–1448.

[51] Yang, Z., J. Ni, and G. Liu, "Delay Compensation of Stretching Signal in OTHR," *6th International Conference on Signal Processing*, Vol. 2, August 26–30, 2002, pp. 1461–1464.

[52] Schikorr, M., "High Range Resolution with Digital Stretch Processing," *2008 IEEE Radar Conference (RADAR '08)*, May 26–30, 2008, pp. 1–6.

[53] Kulpa, K. S., and J. Misiurewicz, "Stretch Processing for Long Integration Time Passive Covert Radar," *2006 International Conference on Radar (CIE '06)*, October 16–19, 2006, pp. 1–4.

[54] Yu, J., et al., "A Single-Chip X-Band Chirp Radar MMIC with Stretch Processing," *2012 IEEE CusTom InTegraTed CircuiTs Conference (CICC)*, September 9–12, 2012, pp. 1–4.

[55] Wojtkiewicz, A., et al., "Use of Polynomial Phase Modeling to FMCW Radar. Part C: Estimation of Target Acceleration in FMCW Radars," *NATO Research and Technology Agency, Sensors & ElecTronics Technology Symposium on Passive and LPI (Low ProbabiliTy of InTercepT) Radio Frequency Sensors*, paper #40C, Warsaw, Poland, April 23–25, 2001.

[56] Kulpa, K., "Target Acceleration Estimation for Continuous Wave Noise Radar," *InTernaTional Radar Symposium 2005*, Berlin, Germany, September 6–8, 2005, pp. 177–183.

[57] Zhang, R., et al., "LFMCW Radar Multi-Target Acceleration and Velocity Estimation Method," *Proceedings 2004 7Th InTernaTional Conference on Signal Processing (ICSP '04)*, Vol. 3, August 31–September 4, 2004, pp. 1989–1992.

[58] Malanowski, M., K. Kulpa, and J. Misiurewicz, "Acceleration Estimation for Passive Coherent Location Radar," *2008 IEEE Radar Conference (RADAR '08)*, May 26–30, 2008, pp. 1–5.

[59] Abatzoglou, T. J., and G. O. Gheen, "Range, Radial Velocity, and Acceleration MLE Using Radar LFM Pulse Train," *IEEE TransacTions on Aerospace and ElecTronic SysTems*, Vol. 34, No. 4, October 1998, pp. 1070–1083.

[60] Xiao, H., W. -D. Hu, and W. -X. Yu, "Parameter Estimation of Accelerating Target Based On Direct Phase Differentiation Method in LFMCW Radar," *International Conference on Information and Automation 2008 (ICIA 2008)*, June 20–23, 2008, pp. 1142–1147.

[61] Smith, S. W., *The Scientist & Engineer's Guide to Digital Signal Processing*, San Diego, CA: California Technical Publications, 1997.

[62] McDonough, R. N., and A. D. Whalen, *Detection of Signals in Noise*, 2nd ed., New York: Academic Press, 1995.

4

Masking Effects

In the previous chapter single-target detection using a noise radar was presented. Typical noise radar processing based on the calculation of the range-Doppler correlation function described by (3.8) [1] is optimal only in the case where a single target echo creates the received signal. In practical cases, however, the radar observes many targets [2], and in addition to target echoes, ground clutter echoes and weather clutter echoes also exist in the received signal. In a pulse radar the echoes reflected from scatterers at different ranges are separated in time. In continuous-wave radars, and especially in noise radars characterized by medium and long integration time, all targets and clutter echoes overlap and interfere with each other. The overlap of echoes leads to effects known as masking effects [2–4]. Weak echoes of far-off targets can be masked by much stronger echoes of nearby targets or ground clutter reflections. In this chapter a detailed description of the masking effect together with algorithms designed to mitigate this problem will be presented [2, 5].

The correlation detection concept described by (3.6) is not optimal when more than one target's echo exists in the received signal. The received signal containing many echoes may be represented by:

$$x_R(t) = \xi_t(t) + \sum_{i=1}^{I} A_i x_T\left(t - \frac{2R_i(t)}{c}\right)\exp\left(2\pi j \frac{-2R_i(t)F}{c}\right) \tag{4.1}$$

For the case of a medium correlation time the received signal can be represented by the formula:

$$x_R(t) = \xi_t(t) + \sum_{i=1}^{I} A_i x_T\left(t - \frac{2R_i}{c}\right)\exp\left(-j2\pi \frac{2v_i F}{c} t\right) \tag{4.2}$$

which is a simplified form of (4.1). The number I stands for the total number of objects observed by the radar, Ri is the range to the ith target, and vi is the ith target velocity. Remember that (4.2) is a simplification of the target echo model valid for medium integration time, where no target range and velocity walk occurs.

The single target optimal detection is based on the concept of matched filtering. While the target range and range velocity are unknown, a bank of filters matched to different target ranges and range velocities is used. The maximal signal is present at the output of the filter tuned to target range and range velocity. In the case where multiple targets echoes exist in the received signal, the matched filter concept is not optimal. In the case where the selected matched filter is tuned to the parameters of one target (range and velocity), the echoes originating from other targets add their power to the power of the receiver noise and increase the processing noise floor [2]. This effect can decrease the detection range in the case where the power of one or more echoes exceeds the receiver's noise power (characterized by the receiver's effective temperature). The detailed description of this effect is a subject of interest in this chapter.

To simplify the problem stated above, let us assume that two target echoes are present in the received signal: one strong echo originating from a nearby target and one weak echo originating from a far target. Detection of the first, strong echo will be performed almost perfectly without any degradation, while detection of the weak echo will be difficult or impossible in practice due to the interference caused by the strong echo. In such a case, when the matched filter is tuned to the far echo, the first, strong echo signal can be treated as additional noise, increasing the receiver noise floor above the thermal noise level to the value:

$$P_n = \frac{P_T G_T S_1 S_R}{(4\pi)^2 r_1^4} + kT_{\mathrm{Re}}B \tag{4.3}$$

where S1 denotes the radar cross-section of the first (nearby) target, r1 denotes the nearby target's distance to the radar, and T_{Re} denotes the effective receiver temperature. The increase of the noise floor causes a decrease of radar sensitivity and consequently degrades the radar detection range. This effect can be treated as masking weak (usually far) targets by strong (nearby) ones. The far-target detection range, calculated by substituting the noise power calculated from (4.3) into the classical range equation, is shortened in such a case to the value:

$$r < r_1 \sqrt[4]{\frac{S_0 t_i B}{D_o S_1} \frac{1}{(1+\gamma)}} \tag{4.4}$$

where $\gamma = \frac{r_1^4 k T_{Re} BL(4\pi)^3}{S_1 P_T G_T G_R}$ is the thermal noise related parameter of a value much smaller than one in the presence of a nearby strong echo. The strong echoes reduce the detection range to the value $r_1 \sqrt[4]{\frac{St_i B}{D_o S_1}}$. This value is sometimes several orders of magnitude smaller than the free-space, single-target detection range. In the case of receiving many strong echoes originating from point-like targets or range and Doppler spread clutter, the situation is still worse. The power of the total interfering noise is a sum of the receiver thermal noise power and the powers of all strong echo signals.

Another approach to the masking problem is to convert the additional noise to the classical loss factor of radar range equation (3.12). The losses calculated using such a concept for classical noise radar are presented in Figure 4.1. The diagram was calculated assuming that the noise radar observed two objects of the same cross-section at different ranges and the echoes of both targets were much stronger than the receiver thermal noise. The time-bandwidth product was 50 dB (100,000). The horizontal axis is the ratio of the far to near distance,

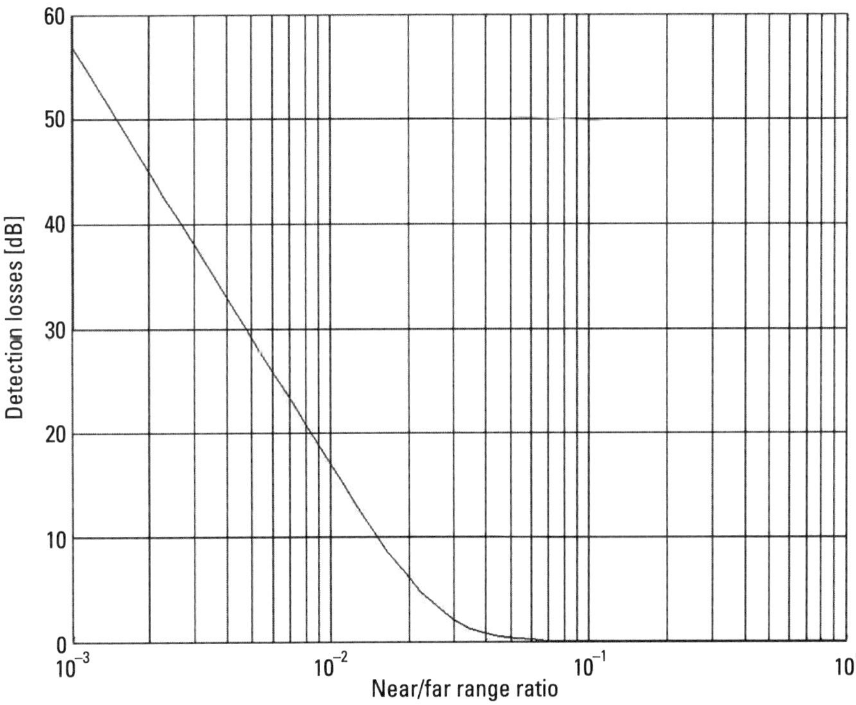

Figure 4.1 Detection losses in the case where two targets of the same radar cross-section at different ranges are observed by noise radar.

and the vertical axis presents the losses in decibels. As one can see, no losses occur only if the far-to-near range ratio is above 0.1 (e.g., the far target is 100 km from the radar and the near target is 10 km away). If that ratio is equal to 0.01 (e.g., the far target is 100 km from the radar and the near target is 1 km away), losses reach 18 dB (the radar detection range is decreased 2.8 times). If such a ratio reaches the value 0.001 (e.g., the far target is 100 km from the radar and the near target is 100m away), losses rise to 58 dB and the detection range is reduced 28 times, under the assumption that both targets have the same radar cross-section.

In practice many radars observe very large ground surfaces in the near distance, so a noise radar equipped only with a range-Doppler correlator will be almost blind due to very strong ground clutter echoes. The masking problem is also illustrated in Figures 4.2 and 4.3.

Figure 4.2 illustrates the basic case where a single object is visible by noise radar because the input signal-to-noise ratio (SNR) is smaller than the time-bandwidth product and the output (after correlation) SNR is higher than the detection threshold. In such a case the target is detected and its position (range and range velocity) can be estimated. The figure is a simplified one, and the constant (asymptotic) value of the sidelobes (or noise residues), as well as the thermal noise, is plotted. They are random variables, and measured (real-life) values are closer to what is plotted in Figures 3.5 to 3.7 than to a straight line.

When the additional strong echo exists in the input signal and the strong signal to weak signal power ratio is higher than the time-bandwidth product (illustrated in Figure 4.3), then the sidelobes (after processing) of the stronger echo will be higher than the weaker echo peak and in such case the weaker echo will be undetected.

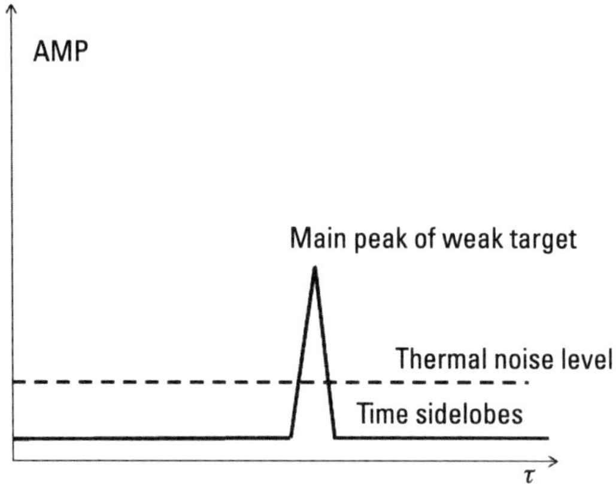

Figure 4.2 The signal components after correlation processing: the main peak of the weak echo is above the thermal noise and thus can be detected.

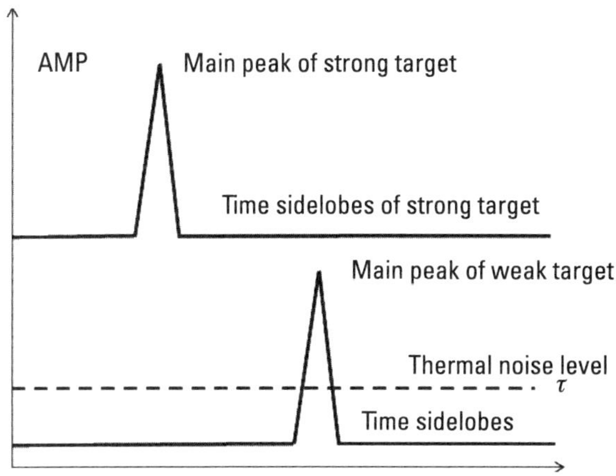

Figure 4.3 An example of the masking effect: signals after correlation processing. The strong target echo is detectable, and the weak target echo is below the sidelobes of the strong target and is not detected.

To avoid the masking effect and restore the original noise radar sensitivity, the received signal must be CLEANED and the strong components have to be removed. The mathematic background and implementation of such processing is presented next.

4.1 Cancellation of the Point-Target Masking Effect

As stated in Chapter 3, classical range-Doppler processing, optimal in the case of a single target, is very sensitive to the presence of high-power components. The noise radar can easily be blinded by the strong echo signal, so to avoid this effect and construct a long-range noise radar robust to the presence of strong ground clutter, additional processing is required.

In modern radar systems all signal processing is performed digitally. The received signal, prior to signal processing, is sampled with the sampling period T_s. It can be represented in the discrete time domain by the formula:

$$x_R(nT_s) = \xi_t(nT_s) + \sum_{i=1}^{I} a_i x_T\left(nT_s - \frac{2R_i}{c}\right)\exp\left(-j2\pi\frac{2v_i F}{c}nT_s\right) \quad (4.5)$$

where n denotes the time instant number and I denotes the number of targets observed by the radar. Assuming that exactly I targets' echoes exist in the received signal, the optimal detection of the targets by noise radar can be achieved by solving the set of nonlinear equations:

$$\left\{\begin{array}{l} x_R(0)-\sum_{i=1}^{I} A_i x_T\left(-\frac{2R_i}{c}\right)=0 \\ \cdots\cdots\cdots\cdots\cdots\cdots \\ x_R(nT_s)-\sum_{i=1}^{I} A_i x_T\left(nT_s-\frac{2R_i}{c}\right)\exp\left(-2\pi j\frac{2v_iF}{c}nT_s\right)=0 \\ \cdots\cdots\cdots\cdots\cdots\cdots \\ x_R((N-1)T_s)-\sum_{i=1}^{I} \begin{array}{c} A_i x_T\left((N-1)T_s-\frac{2r_i}{c}\right) \\ \exp\left(-j2\pi\frac{2v_iF}{c}(N-1)T_s\right)\end{array}=0 \end{array}\right\} \tag{4.6}$$

and finding the unknown values of target echo amplitude vector $\{\hat{A}_i\}$, range vector $\{\hat{R}_i\}$, and velocity vector $\{\hat{v}_i\}$. The number of unknown independent variables is equal to 3I. If the signal model (4.5) is to be free of the receiver noise, then (4.6) could be solved using the block of samples of length N = 3I. In the presence of receiver noise such a solution would be very inaccurate. Better accuracy can be obtained by using a much longer data block. While the number of equations in this case is greater than the number of unknown parameters, the overdetermined equation set can be solved by minimizing the mean-square error. Such an approach is computationally ineffective and sometimes gives false alarms (false detections), when the true number of targets I is unknown.

A more computationally effective approach is based on the adaptive removal of the strong echo components of the received signal [2, 6]. In its simplified form it is assumed that all visible targets can be modeled by point scatterers separated in the range-Doppler domain. If this assumption is not fulfilled, a more elaborated approach, presented in Section 4.3, should be applied.

The initial step of the procedure is the calculation of the range-Doppler cross-correlation function described by (3.8). In the second step the strongest echo is located by finding the maximum of the range-Doppler cross-correlation function. The coordinates of the maximum of the computed absolute value of the range-Doppler cross-correlation function indicate the estimates of target range and velocity. These estimates of the strongest target position $\hat{R}_1$, velocity $\hat{v}_1$, and amplitude $\hat{A}_1$ are further used for modeling the strongest target echo signal. The modeled signal is subtracted from the received signal according to following formula:

$$x_R^{(1)}(t) = x_R(t) - \hat{A}_1 x_T\left(t - \frac{\hat{R}_1}{c}\right)\exp\left(-2\pi j \frac{F_t \hat{v}_1}{c} t\right) \tag{4.7}$$

The signal $x_R^{(1)}(t)$ computed using (4.7) is now free of the strongest target echo signal. The algorithm described above is a modification of a CLEAN type algorithm [7, 8] originally used in radio-astronomy to reduce the problem of observing weak radio sources in the presence of strong ones.

The resulting signal vector $x_R^{(1)}$, which consists of a noise signal and weaker echoes, can be further processed according to the same adaptive echo cancellation procedure until all echoes are removed from the signal, and the noise floor is reached. The method above is based on computations in the time domain. Its main disadvantage is that the estimated target range time delay is usually not the multiplicity of the sampling interval. The estimated target time delay is usually fractional, lying somewhere between the correlation function samples, as shown in Figure 4.4.

As a consequence, a resampling of the reference signal is required. The quality of the cancellation procedure depends significantly on the quality of resampling. To obtain good cancellation of the strong echo signal, a complicated resampling based on high-order polynomial modeling is required. An alternative approach is based on strong echo cancellation in the frequency domain.

Let us assume that both signals (reference and measurement signals) are amplified and downconverted by analog circuits with the same amplitude and phase transfer function $K(\omega)$. After downconversion, the signal is sampled

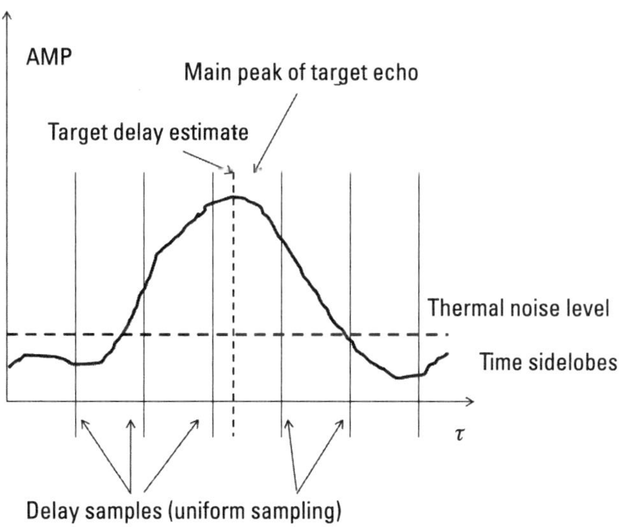

Figure 4.4 Estimation of target range position.

(converted to the discrete time domain). It is also assumed that sampling interval T_s fulfills the Nyquist theorem. In that case the range-Doppler cross-correlation function can be computed digitally according to the formula:

$$y_{l,m} = \sum_{i=0}^{M_i-1} x_R(i) \cdot x_T^*(i-l) \cdot \exp\left(2\pi j \frac{im}{M_i}\right) \tag{4.8}$$

where $x_R(i)$ denotes the ith sample of the measurement signal, $x_T(i)$ denotes the ith sample of the reference signal, i denotes the sample index, l denotes the time delay index (corresponding to the range $l\frac{T_s c}{2}$), m is the Doppler shift index (corresponding to the range velocity), and Mi is the integration length (corresponding to the integration time $M_i Ts$). To obtain higher computational efficiency, the fast Fourier transform (FFT) is often used to calculate the cross-correlation (4.8). Let $X_R = \text{FFT}(x_R(0..(M_i - 1)))$ be a vector containing the Fourier transform of the measurement signal vector of length M_i, and Let $X_T = \text{FFT}(x_T(0..(M_i - 1)))$ denote a vector containing the Fourier transform of the reference signal. Additionally, let us denote a circular shift function SHIFT(X, m) that returns the vector X shifted by m elements clockwise. A single row of the range-Doppler function (4.8), for a selected Doppler shift index m, can be treated as the inverse Fourier transform of the product of the Fourier transform of the measurement signal (usually zero-padded) and the complex conjugation of the shifted Fourier transform of the reference signal (also usually zero padded), that is,

$$y_{l,m} = FFT^{-1}\left(X_R SHIFT\left(X_T^*, m\right)\right) \tag{4.9}$$

The location of the maximum value of $y_{l,m}$ can be found in the discrete domain l,m. The discrete estimates of the Doppler frequency and range of the target has insufficient accuracy for strong echo cancellation described by (4.7), and more accurate estimates of those parameters should be used. The simple estimation method based on fitting a parabolic curve to the logarithm of the absolute value of $y_{l,m}$, described by the author in [9], has been selected for modeling the main peak of the shape and for estimating the target range and velocity by finding the coordinates of the maximum value of the parabolic curve, both in range and Doppler. This method has been used for all simulations presented in this chapter.

The estimated values of Doppler shift $\hat{m}_1$ and range shifts $\hat{l}_1$ are not integer numbers required for a direct use of the discrete time version of (4.7), because the time shift is usually not an integer multiple of the sampling time T_s. To overcome this problem, reference signal resampling is required for modeling the strong echo signal in the time domain. In the frequency domain, however, the time resampling is not needed [2]. In this domain the well-known time-shifting properties of the FFT transform, $FFT(x(t-\tau)) = FFT(x(t))e^{-j\tau\omega}$, can be used. The strong echo signal model in the frequency domain can be evaluated by computing the Fourier transform of the reference signal vector multiplied element by element with a harmonic signal of the estimated Doppler frequency $\hat{m}_1 / M_i$ (frequency shifting of the reference signal). The computed FFT vector has to be multiplied (element by element) by a harmonic signal of frequency $\hat{l}_1 / M_i$ (which is equivalent to the time shifting of the reference signal by $\hat{l}_1 / M_i$):

$$X_{R1} = FFT\left(\left\{ {}_i x_T(i) e^{2\pi j \frac{i\hat{m}_1}{M_i}} \right\}\right) \left\{ {}_n e^{2\pi j \frac{n\hat{l}_1}{M_i}} \right\} \tag{4.10}$$

where $\{i^*\}$ is a vector indexed by i. The complex amplitude of the strong echo signal can be computed as a partial correlation coefficient [10, 11] between the received signal and the estimated normalized strong echo signal in the frequency domain:

$$A_{R1} = \frac{X_R X_{R1}^*}{|X_R|^2} \tag{4.11}$$

Strong echo cancellation can be carried out in the frequency domain by subtracting the product of the estimated complex amplitude and the normalized strong echo estimates from the received signal. The cancellation is expressed by the following formula:

$$X_R^1 = X_R - A_{R1} X_{R1} \tag{4.12}$$

The resulting output signal (in the frequency domain) is further subject to the detection process of remaining weaker echoes. These computations can be performed iteratively until all strong echoes are removed from the received measurement signal. The procedure described above requires $L + 1$ calculations of the range-Doppler cross-correlation function, where L is the number of iterations.

The question is how many dominant scatterers it is necessary to remove, and whether it is possible to terminate the procedure before removing the last echo.

The answer is positive. It is possible to terminate the procedure when the strongest echo is much smaller (say, more than 10 dB) than the noise floor plus time-bandwidth product (in the decibel scale). If such a condition is reached, then all the next targets will have a rather small contribution to the noise floor (except in the case where there are many of them).

Such an approach, even if terminated earlier, is very ineffective from the point of view of the computational power requirement. However, it is possible to decrease the computational complexity by calculating the range-Doppler cross-correlation function only twice (or a few times). In the first stage, the range-Doppler cross-correlation function is calculated using the received signal. In the second stage, all strong targets are detected and their range and Doppler coordinates are estimated. In the third stage, the strong target echo removal procedures are carried out for all detected targets, sorted from strongest to weakest. In the fourth stage, the final range-Doppler cross-correlation function is calculated and the weak targets are detected. Such a procedure can introduce some detection losses and produce some false alarms, but it requires much less computational power. If the total power of all the echoes detected in the fourth stage is smaller than the thermal noise then the procedure should be terminated. If not, then the cleaning and detection stages should be repeated once more.

The example results obtained using the algorithm described above are presented in Figures 4.5 through 4.7. For the simulation the signal bandwidth of 100 kHz was selected and the integration time was equal to 0.1 second. The reference signal was modeled as band-limited Gaussian noise obtained by the lowpass filtering of complex white noise. The thermal noise was modeled as complex Gaussian white noise. The measurement signal consisted of two echoes of different amplitudes, shifted both in range (time shift) and velocity (Doppler shift). The S/N ratio for the first, strong echo was equal to 20 dB (calculated at the input, before correlation processing), the range was 5 km, and the velocity was −300 m/s. The second, weak signal had an S/N equal to −20 dB, a range of 31 km and a velocity of 150 m/s. The time-bandwidth product was equal to 10,000 (40 dB).

The range-Doppler cross-correlation function calculated for the measurement and the reference signals is presented in Figure 4.5. The strong echo peak is clearly visible while the weaker signal is completely buried (not visible) in the correlation sidelobes (noise floor). To obtain the detection of the weak echo, the CLEANing procedure was applied. The result of the strong echo cancellation procedure is presented in Figure 4.6. After applying the cancellation procedure, the strong echo has been suppressed more than 40 dB (together with the corresponding sidelobes) and, due to this, a weak echo peak can easily be detected.

For a comparison Figure 4.7 presents the magnitude of the range-Doppler cross-correlation function in the case where the measurement signal consists only of the weaker echo and the thermal noise. There is only a slight difference in the noise floor sidelobe structure, but the noise floor takes the same level in both cases. In the experiments presented above, the reference signal was a

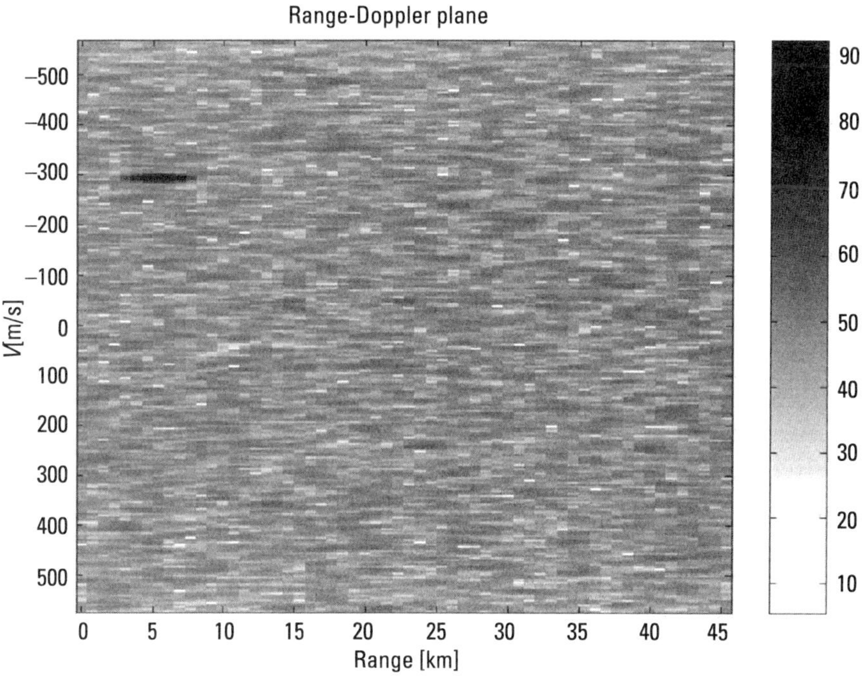

Figure 4.5 The magnitude of the range-Doppler cross-correlation function. Two moving targets echoes are present in the received signal: a strong one and a weak one. A weak echo has been masked by correlation sidelobes. Grayscale is in decibels.

"pure" one, which means that it was free of any distortion and additional noise components (such as reference channel receiver noise). In real situations, however, the reference signal also includes a white noise component weaker than the reference signal by a factor of the S/N ratio of the reference channel (which can be at a level of 60 to 110 dB, depending on the quality of the mixer and A/D converter used). The noise of the reference channel receiver is then added to the measurement channel signal during the strong-echo-removal procedure. As a result, the performance of the presented method is limited by the reference channel's S/N ratio.

The procedure described above is effective only for mitigating the masking effect of point-like target echoes. This procedure can be easily applied in low and medium resolution radars for ensuring radar sensitivity in the direction of strong target echoes. It is also possible to implement the strong echo cancellation procedure in the long time integration radar. Radars of this type require the additional stretch processing of the reference signal, described in Section 3.4.1. For range spread target echoes spanning several resolution cells, it is necessary to apply the more advanced method presented in the next section.

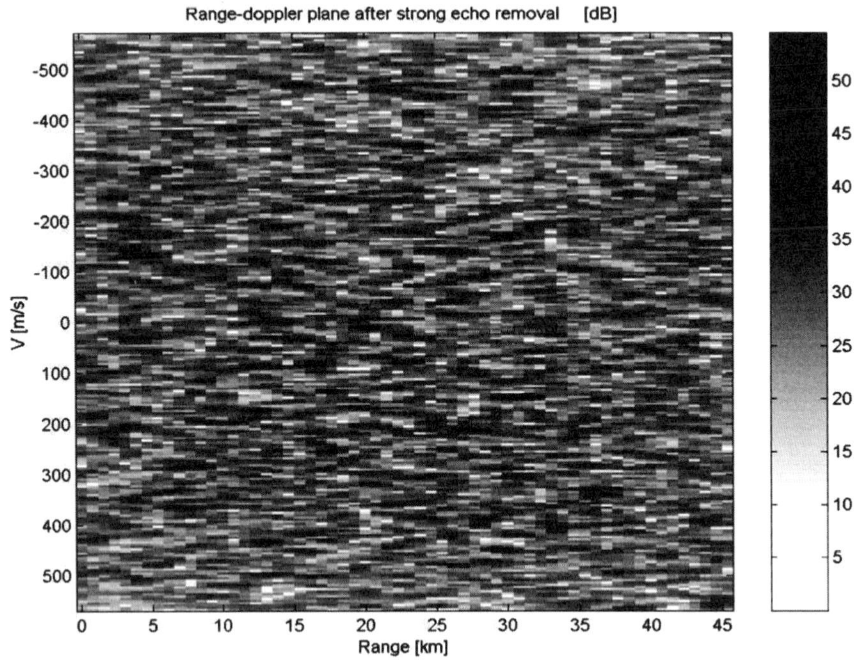

Figure 4.6 The magnitude of the range-Doppler cross-correlation function after strong echo cancellation. The strong echo is attenuated by 43 dB, the weak echo is visible, and the processing noise floor (correlation sidelobes) is attenuated by 25 dB. Grayscale is in decibels.

4.2 The Stretch Processing for Point-Like Target Cancellation

In the previous chapter a very simple target echo model was applied. It was assumed that the time delay is constant and the target movement only influences the Doppler frequency, which was also constant during observation time. Now we will add another effect to the signal model: a time stretch under the assumption of a constant target radial velocity v. Under such an assumption, the range to the target can be expressed as $R(t) = R_o + v_t$. In (4.2) the influence of the target velocity is limited only to the Doppler effect, which is a good approximation when the observation time is short enough. For a longer integration time the received baseband signal differs from the transmitted one by the Doppler modulation of the phase and by its envelope being delayed and time-scaled:

$$x_R(t) = Ax_T\left(t - \frac{2R + 2vt}{c}\right) \cdot \exp\left(-4\pi jvt\, F/c\right) \tag{4.13}$$

In (4.2) the factor $2v/c$ in the envelope argument was assumed negligible with respect to one, which is not true for the long integration time. Therefore,

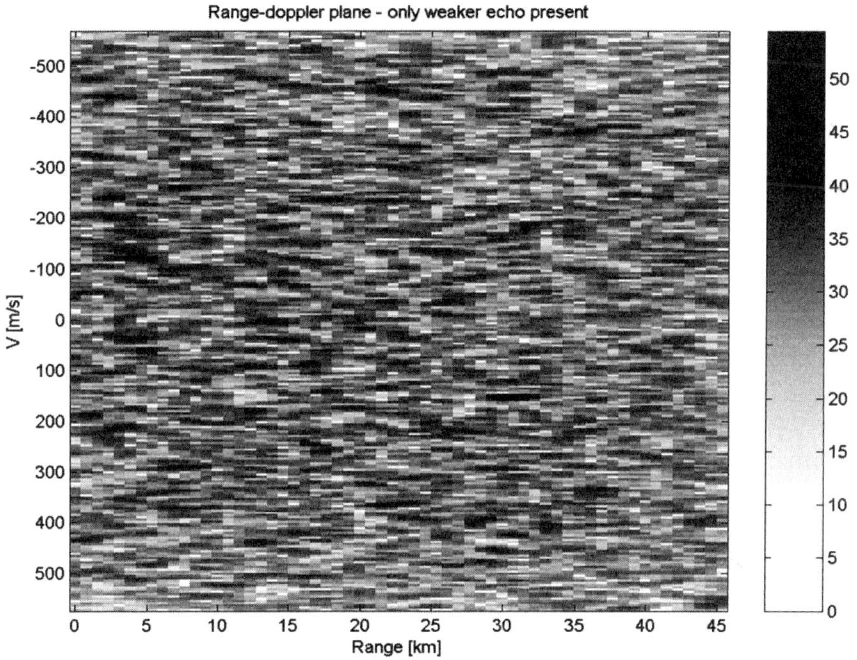

Figure 4.7 The magnitude of the range-Doppler cross-correlation function in the case where only the weak echo is present in the received signal. Grayscale is in decibels.

to calculate the long-time model of the signal it is necessary not only to shift the transmitted signal envelope in time, but also to perform the time stretch operation. For longer integration times more complicated target motion models should also be taken into consideration. For the constant acceleration in $R(t)$, the range to the target can be expressed as $R(t) = R_o + vt + \frac{at^2}{2}$. In this case the target echo signal can be modeled as:

$$x_R(t) = Ax_T\left(t - \frac{2R + 2vt + at^2}{c}\right) \cdot \exp\left(j2\pi(2vt + at^2)F/c\right) \quad (4.14)$$

In this model the Doppler component has a polynomial phase, which makes the correlation processing more complicated due to the necessity to estimate the phase polynomial coefficients. A detailed description of this problem can be found in the previous chapter.

If the time of the observation is extended even further, higher derivatives of range $R(t)$ should be taken into consideration, and also the amplitude must be treated as a variable of time, as the received power is related to the radar-target distance and the aspect ratio. Also in such a case the target rotational motion can influence the model, while target rotation can introduce the amplitude modulation due to the variation of RCS versus aspect angle.

The quality of the CLEANing procedure depends on the quality of the model used (agreement between the model and the echo signal) and the quality of model parameter estimation. In the case where the range or Doppler position of the target is estimated with a large error, the CLEANing procedure will cancel only part of the signal and the echo residue power will remain.

Figure 4.8 presents the echo residual power after the removal of a strong object versus a normalized target range estimation error (range error to range resolution cell size). The 0-dB level in the residual power represents the power of the strong object and means that residual power is equal to echo power, so the cancellation was not successful at all. It can be seen that a 10% range estimation error will limit the quality of the CLEAN procedure to the value of about 14 dB and a 1% accuracy to the value of 34 dB, and, to obtain more than 60 dB, of CLEANing, it is necessary to have a range accuracy better than 0.1%.

A similar analysis, but in this case for velocity accuracy, is presented in Figure 4.9. The results are almost identical to the range case and a 10% Doppler frequency (range velocity) estimation error will result in the limiting of the CLEAN procedure to the value of about 14 dB and a 1% accuracy to the value

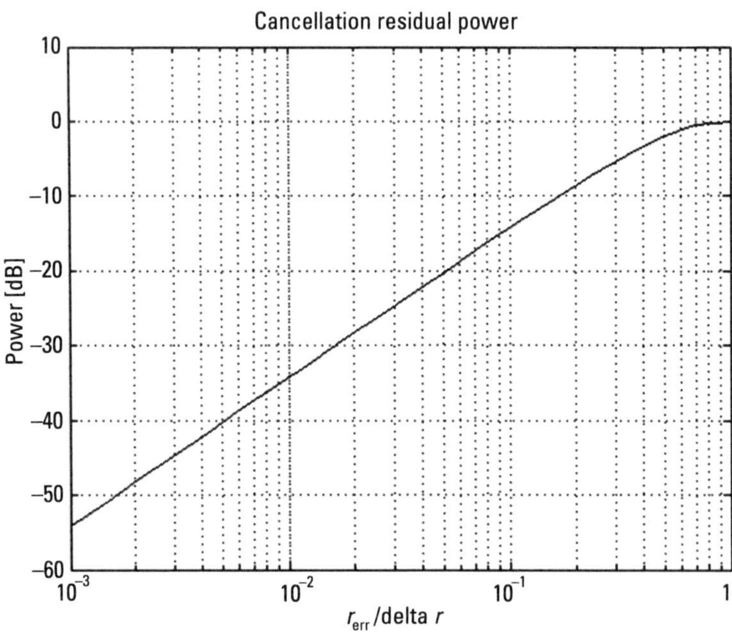

Figure 4.8 Residual power after the CLEANing procedure versus range error, normalized to the range resolution cell.

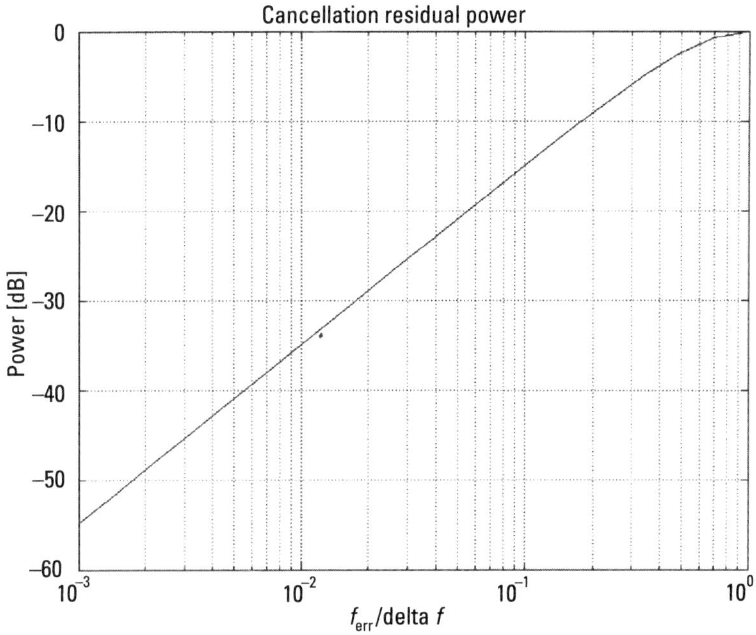

Figure 4.9 Residual power after the CLEANing procedure versus the Doppler frequency error, normalized to the Doppler frequency resolution cell.

of 34 dB, and, to obtain more than 60 dB of CLEANing, it is necessary to have a velocity accuracy of better than 0.1%.

Another factor is the range migration of the target during coherent integration time. Such migration implies two effects in noise radar. The first effect implies integration losses while the time the target remains in the resolution cell is shorter than the integration time. The second implies losses in the CLEANing procedure, while the modeled signal differs from the received one.

The losses in the correlation receiver due to the neglecting of the stretch effect are shown in Figure 4.10. The losses are defined here as the degradation of the S/N at the correlation receiver output for a moving target with respect to a stationary target of the same echo strength. The x-axis is scaled with vt/dR, the target migration during observation (integration) time t, normalized to the radar resolution cell size in range dR. As can be seen from the plot, the losses start to be visible when the range migration in the integration time approaches the range resolution cell size and becomes very high for a migration of several range cells.

An example of the second effect is presented in Figure 4.11. The residual power after the removal of a strong object is shown versus the target migration normalized to the range resolution cell. It can be seen that the effective removal of the object echo becomes a problem much earlier than when the correlation

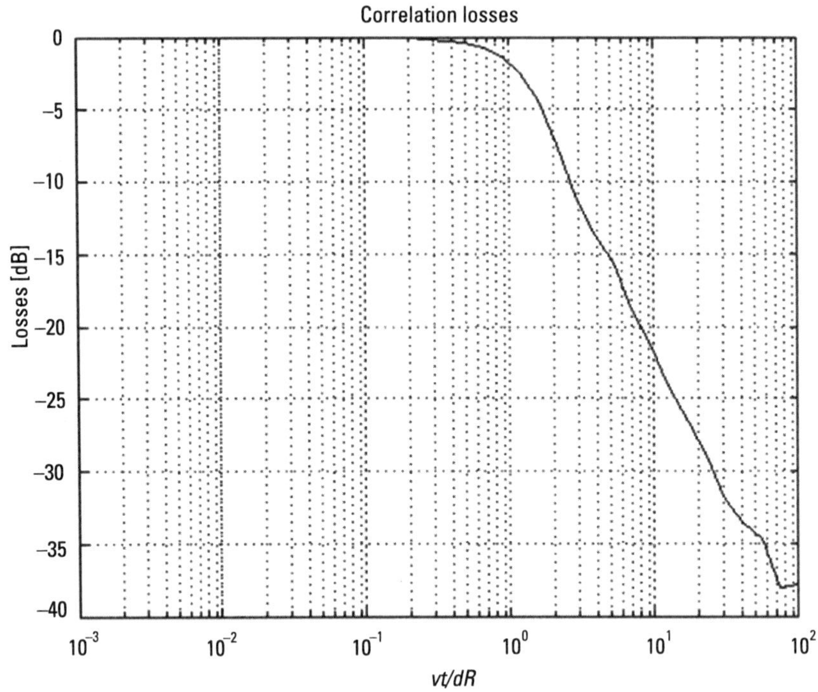

Figure 4.10 Correlation losses versus target range walk normalized to the range resolution cell.

losses show up. If the target migration is within 1% of the resolution cell, then it is possible to clean the target with 44-dB quality, which is sufficient for most cases. However, such a small migration is only for slow targets, with a velocity equal to 1% of the maximum assumed target velocity (10 m/s for $v_{max} = 1{,}000$ m/s). The echo of the targets that migrate 10% of the range resolution cell will be attenuated only by 24 dB, and targets that migrate through the entire resolution cell will be attenuated only by 4 dB.

The avoidance of such effects is possible with stretch processing. In the derivations of both the correlation receiver and the strong echo canceler, the expected target echo is modeled using the known transmitted signal (a template) as a base. In the case of stretch processing, the construction of a modeled echo [according to (4.13)] requires modulating the baseband template with the target's Doppler frequency (to match the Doppler shift in the received signal) and stretching the template in time (to match the envelope stretch).

In practice all the processing is done by digital means. The transmitted signal template is obtained either by the sampling of an analog noise or just by storing the digitally generated pseudo-noise signal that is used to modulate the carrier. Thus, the stretching of the reference signal requires the changing of the sample rate of the signal according to the equation:

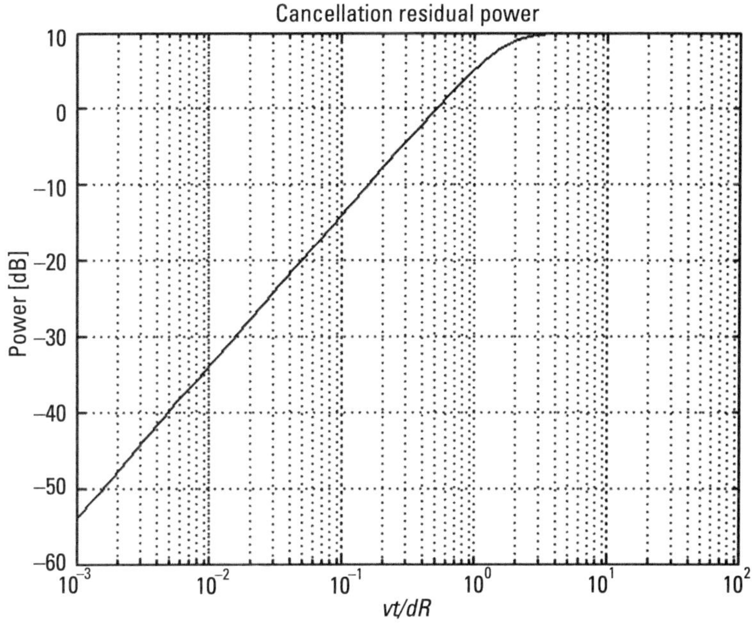

Figure 4.11 Residual power after the cancellation of a strong object versus range migration normalized to the range resolution cell.

$$x_{ref}(n,v) = x_T\left(nt_s \cdot \left(1 - \frac{2v}{c}\right)\right) \tag{4.15}$$

where $x_T(t)$ is the continuous representation of a discrete-time transmitted signal template $x_T(n)$ and t_s is the sampling interval. Depending on the target direction, the velocity v can be positive or negative; the time scale has to be dilated or compressed.

As it can be seen from the above equation, with the practical values of velocity v the sample rate ratio differs from 1 by a very tiny fraction, for example, for the $v = 150$ m/s the ratio $1 + 2v/c$ is equal to 1 ± 10^{-6}. This makes the effective resampling a complicated task to be done in real time. Many of the popular methods developed (e.g., for image processing) are suitable only for rescaling by a ratio of two relatively small integers. Several fast methods will be discussed next.

One of the simplest resampling methods, used mainly in image processing, is the nearest-neighbor interpolation [13]. The method is very fast, but at the cost of poor accuracy. In the case of a sample rate change very close to 1 as expressed by (4.15), the nearest-neighbor resampling consists of periodically removing (or duplicating) one of $c/2v$ samples.

Another simple method is the linear interpolation method, where a new sample is calculated as a linear combination of adjacent original samples:

$$x_{ref}(n,v) = \alpha x_T\left(\left\lfloor nt_s \cdot \left(1+\frac{2v}{c}\right)\right\rfloor\right) + (1-\alpha)x_T\left(\left\lceil nt_s \cdot \left(1+\frac{2v}{c}\right)\right\rceil\right) \quad (4.16)$$

where $\alpha = \left\lceil nt_s \cdot \left(1+\frac{2v}{c}\right)\right\rceil - nt_s \cdot \left(1+\frac{2v}{c}\right)$ (the symbols $\lfloor\cdot\rfloor$ and $\lceil\cdot\rceil$ represent rounding down or up to the nearest integer).

Polyphase filtering methods are a computationally efficient form of upsampling by an integer factor N and then downsampling by another factor M, with lowpass filtering to remove aliasing [14]. In a standard application they allow the obtaining of a rational M/N stretch factor using a linear filter with M coefficient sets, changing for each output sample. For a noise radar application, the number of coefficient sets is very large, as the M/N has to approximate a number very close to 1, but a simplified approach using the technique similar to the one described in [15, 16] can result in a smaller number of coefficient sets. First, a simple polyphase resampler is used to calculate the sample values for time instants from a set of points oversampled by some integer factor. Then a simple linear interpolation is performed to find a more exact value at a desired point between these samples. Such an approach will be referred to in the following as the modified polyphase method.

Spectrum-based methods consist of the conversion of a signal to the spectral representation and making the resampling a manipulation in the spectral domain. The manipulated spectrum is then converted back to the time domain (e.g., using an efficient IFFT algorithm). Similar to the polyphase filter resampling, the methods based on FFT allow resampling by a rational factor N/M, where M and N are orders (numbers of points) in FFT and IFFT. The existence of very efficient libraries of functions for the software implementation of the arbitrary order of FFT/IFFT, using state-of-the-art mathematical concepts and programming techniques [17], makes the FFT method a very good choice if the resampling factor can be expressed as a ratio of two integers. In other cases, another spectral method may be of use, namely, chirp-z transform resampling.

The chirp-z transform [14] is an efficient method to compute the z-transform on a set of points that are defined as a^n where $n = N_0 \ldots N_1$. By setting $a = e^{j2\pi(1+2v/c)/N}$ and calculating the chirp-z transform of an N-point FFT, we effectively obtain the signal resampled with exactly the required rate [18]. An example of this method being applied to image scaling is shown in [19]. The efficient implementation of the chirp-z transform is about three times more complex than the FFT. It has been shown that in the case of correlation processing for a PCL radar the simplest interpolation methods perform well [18].

To illustrate the accuracy of the methods described above the result of simulations are presented in Figures 4.12 and 4.13. As the transmit signal, a noise signal with a bandwidth equal to 0.6 of the sampling frequency was used. Three stretching methods were investigated: linear interpolation, modified polyphase (with 2× upsampling followed with linear interpolation), and the chirp-*z* based method. The plot shown in Figure 4.12 confirms the usability of all stretching methods for correlation processing in a noise radar (the losses being within 1 dB). However, only the chirp transform does not introduce losses. It is worth noting that, due to linear resampling errors, the first method introduces losses for small range walks (much smaller than one range resolution cell) while applying no stretch provides better results.

In the case of strong echo cancellation algorithms, the requirements on stretch quality are higher. The results plotted in Figure 4.13 show the cancellation residual power using different resampling methods for stretch processing. In comparison to no stretch processing the linear interpolation method shows some improvement, about 14 dB, which is not enough in a typical application. The improvement occurs only for target range walks similar or bigger than the range resolution cell. For small walks (below 30% of the range resolution cell), there is no improvement. The modified polyphase method performs a little bit better, providing about 21-dB improvement for a long walk and a few decibels for a small walk. Only the chirp-based method allows the attainment of

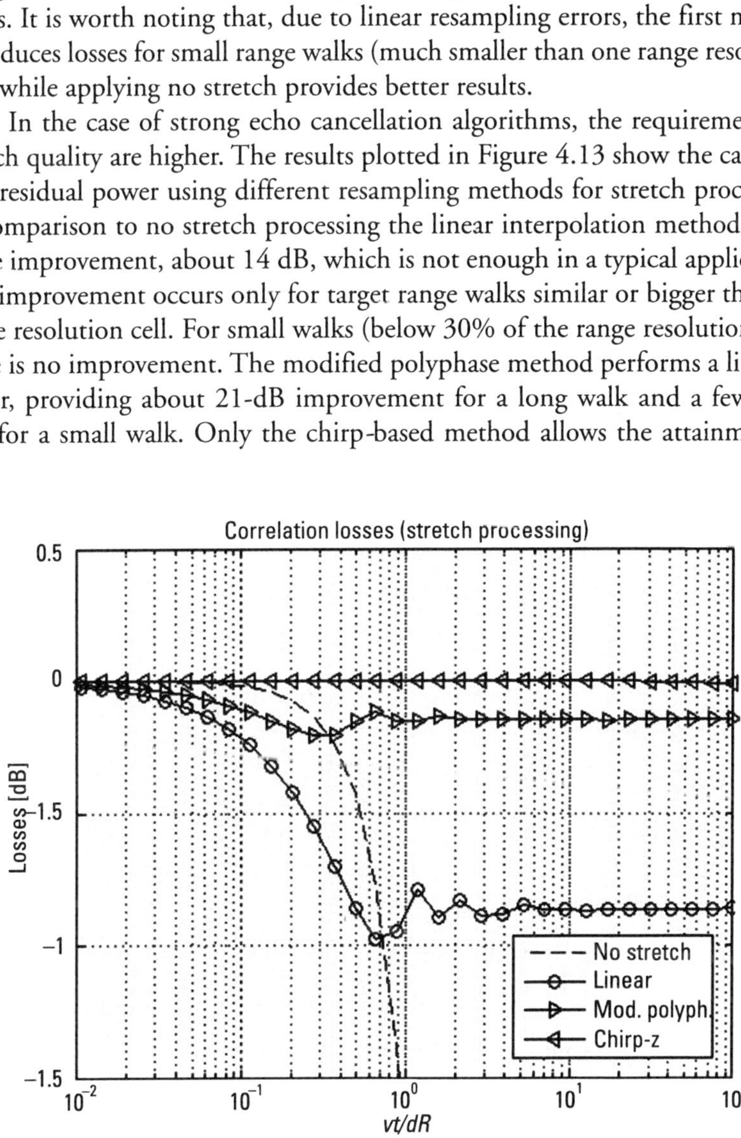

Figure 4.12 Correlation losses after stretch processing.

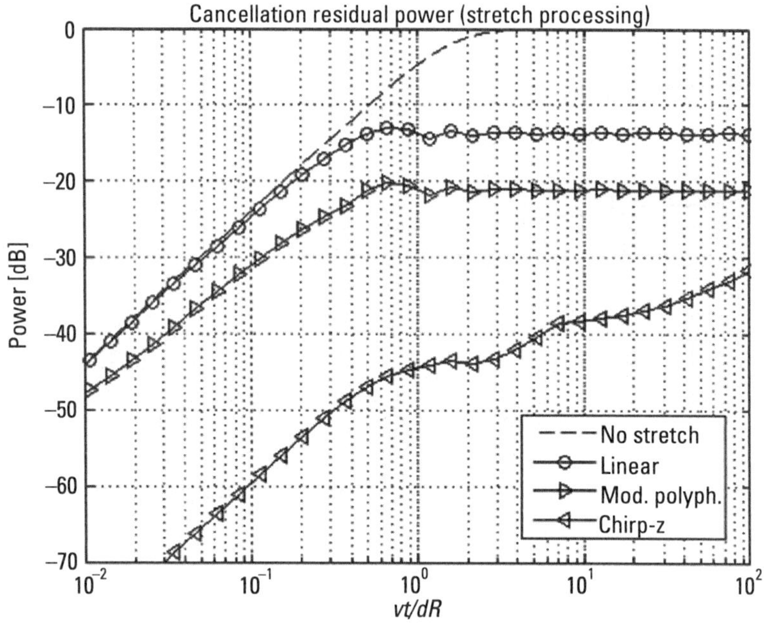

Figure 4.13 Residual power after the cancellation of a strong object with stretch processing.

the desired improvement, about 40 dB over almost all the evaluated migration values range. In most cases this is enough to remove a strong target completely.

4.3 Cancellation of Direct Signal and Range Spread Clutter

In the previous chapter the effective method for the cancellation of the masking effect regarding the point-like targets was presented. In many practical cases the dominant part of masking effects comes from range-spread ground clutter echoes and cross-talk between the transmitting and the receiving antenna. Cancellation of the direct signal, as well as the cancellation of range spread stationary target echoes (ground clutter) spanning several range resolution cells, requires another approach. From one point of view, this part of the process is much more complicated, as range spread clutter has to be attenuated with a much more complicated model, but it is simpler as it deals only with the time shift of the signal while the Doppler shift is zero (or almost zero).

4.3.1 The Simple Method of Direct Signal Cancellation

In an ideal case the separation between the transmitting and the receiving antenna is infinite, so no direct (transmitting) signal is defused to the receiving signal. In real cases separation between these two antennas is limited to the value of tens of decibels. The presence of a transmitted (direct) signal in the received signal produces a high level of sidelobes in the correlation process, which may completely mask almost all echoes of interest. The cross-talk can be treated as the presence of a very strong object very close to the radar, and, using (4.4), it is possible to calculate the detection range in the case of cross-talk.

The power of the received target echo signal is described by (3.7). The power of direct (crosstalk) signal at the receiver input is equal to:

$$P_d = \frac{P_T C}{L} \tag{4.17}$$

where PT denotes transmitted signal power, C denotes cross-talk coefficient, and L stands for radar (transmitting and receiving) losses. Usually the direct (cross-talk) signal power is much higher than the receiver thermal noise power. The target echo may be detected if its power is strong enough and fulfills the constraint:

$$P_R \geq P_d \cdot D \tag{4.18}$$

Substituting (4.17) and (3.7) into the above inequality, we obtain:

$$\frac{P_T G_T G_R S_o \lambda^2}{(4\pi)^3 L R^4} \geq \frac{P_T C}{L} \frac{D_0}{t_i B} \tag{4.19}$$

It follows from relation (4.19) that the maximum detection range is equal to:

$$R_{\max} = \sqrt[4]{\frac{G_T G_R S_o \lambda^2 t_i B}{(4\pi)^3 D_0 C}} \tag{4.20}$$

For example, the maximum detection range for a 1-m^2 target for a radar working in the X-band (carrier frequency 10 GHz) with the processing gain of $t_i B = 50$ dB, an omnidirectional transmitting antenna, a directional receiving antenna with a 30-dB gain and –80-dB crosstalk between antennas, is equal to 120m. This detection range is independent of the transmitting signal power (assuming that this power is high enough to produce cross-talk significantly higher than the receiver thermal noise power, in this case higher than a few microwatts). To obtain the detection range of 40 km, the additional 100-dB direct signal attenuation is required. The numbers presented show that to construct a useful noise radar it is necessary to ensure a very high dynamic range and very low cross-talk between the transmit and receive antennas.

The direct signal can be attenuated using a simple adaptive procedure. Let us assume that the received signal is in the form:

$$x_R(t) = \xi_t(t) + A_d x_T(t) + \sum_{i=1}^{I} A_i x_T\left(t - \frac{2R_i}{c}\right)\exp\left(-j2\pi\frac{2v_i Ft}{c}\right) \quad (4.21)$$

where A_d denotes the complex amplitude of the direct signal. To attenuate the direct signal, it is necessary to estimate the direct signal amplitude $\hat{A}_d$ and to subtract the amplitude weighted reference signal from the received (measurement) one. Similarly, as in the previous case, it is possible using N discrete time measurements to form the overestimated equation set with a single unknown A_d:

$$\left\{\begin{bmatrix} x_R(0) = A_d x_T(0) \\ x_R(T_s) = A_d x_T(T_s) \\ \ldots\ldots\ldots\ldots\ldots \\ x_R\left((N-1)T_s\right) = A_d x_T\left((N-1)T_s\right) \end{bmatrix}\right\} \quad (4.22)$$

and solve this equation set by minimizing the mean square error. Let $\{x_R\}$ and $\{x_T\}$ be column vectors of the received and transmitted (reference) signal, respectively. According to the linear algebra method [20], the mean square estimate of the direct signal amplitude $\hat{A}_d$ can be calculated as follows:

$$\hat{A}_d = \{x_T\}'\{x_R\} \big/ \{x_T\}'\{x_R\} \quad (4.23)$$

where * stands for vector transposition (with complex conjunction). The solution (4.23) can also be rewritten in the form:

$$\hat{A}_d = R_{x_T x_R}(0) \big/ R_{x_T}(0) \quad (4.24)$$

where $R_{x_T x_R}(0)$ is the zero lag of the cross-correlation function between the transmitting and receiving signals and $R_{x_T}(0)$ is the zero lag of the autocorrelation function of the transmitted signal. After applying the direct signal removal procedure, the measurement signal $x_R^{(d)}(t)$ can be written as:

$$x_R^{(d)}(t) = x_R(t) - \hat{A}_d x_T(t) \quad (4.25)$$

In practical cases the direct signal is not removed completely. The direct signal attenuation factor depends on several factors. The most important factor is the colinearity of phase and the amplitude transmission characteristic $K(\omega)$ of the surveillance and reference channels. The absolute value of the gains of those two channels are not important while the actual cross-talk is estimated and removed adaptively, but the mismatch in the frequency characteristic decreases the direct signal attenuation factor. It is very difficult to ensure the required colinearity of channels by using analog filters, so it is recommended to use wideband analog front-end and digital filters for final signal filtering. The second factor, also very important, is the time delay τ_C between the point from which the reference signal is acquired and the point of cross-talk. For the relatively small time delay τ_C the receiving signal can be represented by:

$$x_R(t) = \xi_t(t) + A_d x_T(t + \tau_C) + \sum_{i=1}^{I} A_i x_T\left(t - \frac{2r_i}{c}\right) e^{j2\pi\left(-\frac{2v_i F}{c}\right)t + \phi_i} \quad (4.26)$$

The use of the procedure described above does not allow the removal of the cross-talk signal fully while there is decorrelation between the reference signal and the cross-talk. Assuming that the transmitted signal has a rectangular form spectrum of bandwidth B, then the autocorrelation function for small delays is very close to the sinc function; that is:

$$R_{X_T}(\tau) \cong \frac{\sin(\pi\tau B)}{\pi\tau B} \quad (4.27)$$

The power of the remaining direct signal can be then calculated according to the formula:

$$P_{d_R} \cong P_d\left(1 - \left(\frac{\sin(\pi\tau B)}{\pi\tau B}\right)^2\right) \quad (4.28)$$

The attenuation factor of the direct signal (P_{d_R}/P_d ratio) versus time delay (normalized by the signal bandwidth) is illustrated in Figure 4.14.

It follows from the plot shown in Figure 4.14 that a good enough attenuation (more than 100 dB) can be obtained only in the case where the time delay between the point when the reference signal is sensed and the point of cross-talk multiplied by signal bandwidth is in the order of 10^{-3}. Assuming a 10-MHz signal bandwidth, the physical distance between those two points must be kept within 3 cm. However, it is usually difficult to fulfill this requirement. Moreover, in many practical cases there are several sources of crosstalk, separated in

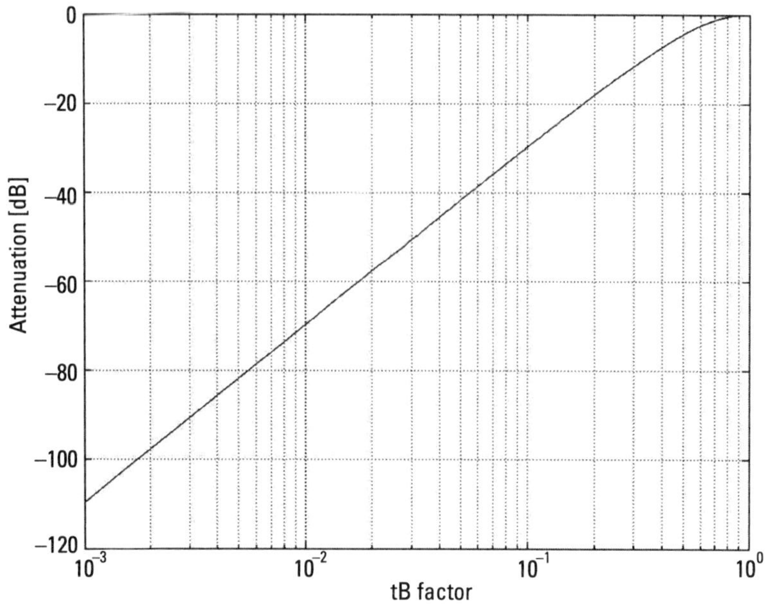

Figure 4.14 The attenuation factor of the direct signal versus time delay between the reference signal and crosstalk normalized by signal bandwidth.

space. In order to obtain a good attenuation of the direct signal it is necessary to use a more elaborate cancellation procedure, which is presented in the next section.

4.3.2 Ground Clutter Cancellation

There is another reason for the decrease of noise radar sensitivity besides the direct signal cross-talk: strong ground clutter echoes coming from the vicinity of the radar can significantly increase the processing noise floor or even saturate a radar receiver. In typical pulse radars two mechanisms of ground clutter mitigation are applied. The first is sensitivity time control (STC), which changes the gain of the radar receiver or introduces additional losses just after the pulse emission in the first tenths of microseconds to avoid receiver saturation by strong nearby echoes. The second is the attenuation of ground clutter by using classical moving target detection or moving target indication (MTD/MTI) techniques. The received signal is passed through the bandpass or high-pass filters that attenuate the zero-Doppler echoes. It was shown in [5] that such an approach is not effective in noise radars, where filter-based removal of ground clutter does not remove the sidelobes of ground clutter and a simple MTD procedure does not eliminate the masking effect.

To remove the ground clutter echo signals together with the sidelobes related to them, a more complicated adaptive removal procedure should be applied [5, 21].

The received signal containing ground clutter echoes (without moving targets) can be represented by:

$$x_R(t) = \xi_t(t) + \int_{r=0}^{r=R_{cm}} A(r) x_T\left(t - \frac{2r}{c}\right) dr \tag{4.29}$$

where $A(r)$ denotes the range distribution of complex amplitudes of ground clutter echoes, and Rcm denotes the maximum range of the ground clutter (limited by Earth geometry and/or by the transmitted signal power). Formula (4.29) can also be used for modeling a direct signal, for which r is close to zero. Assuming that the spectrum of the received signal is band-limited, it is possible to apply time sampling with sampling period $T_s < 1/B$. The sampler produces the discrete time signal described by the following formula:

$$x_R(nT_s) = \xi_t(nT_s) + \int_{r=0}^{r=r_{cm}} A(r) x_T\left(nT_s - \frac{2r}{c}\right) dr \tag{4.30}$$

In the case where the change rate of the clutter reflectivity is small, then the sampling theorem can also be applied in the spatial domain. Let us assume that the point-like clutter equivalent to (4.30) is distributed in equally distant points [5]:

$$r_i = \frac{iT_s}{2c} \tag{4.31}$$

On this assumption the model of clutter signal can then be rewritten to the form:

$$x_R(nT_s) = \xi_t(nT_s) + \sum_{i=0}^{I} A_i x_T((n-i)T_s) \tag{4.32}$$

It was shown that this model is also useful for predicting the clutter signal without the assumption of the slow change rate of clutter reflectivity.

To remove the ground clutter and the direct signal, it is necessary to estimate the unknown vector of the complex amplitudes $[A_i]$ of the cross-talk signal and clutter echoes and to subtract the modeled clutter signal from the received one. The unknown amplitude vector $[A_i]$ can be estimated by solving the equation set:

$$\left\{\begin{array}{l} x_R(0)=\sum_{i=0}^{I-1} A_i x_T\left((-i)T_s\right) \\ \ldots\ldots\ldots\ldots\ldots \\ x_R(nT_s)=\sum_{i=0}^{I-1} A_i x_T\left((n-i)T_s\right) \\ \ldots\ldots\ldots\ldots \\ x_R\left((N-1)T_s\right)=\sum_{i=0}^{I-1} A_i x_T\left((N-1-i)T_s\right) \end{array}\right\} \tag{4.33}$$

The number of unknown independent variables is equal to I, and thus the minimal number of equations is $N = I$. In the absence of the receiver thermal noise, this number of equations would be adequate and the estimated vector $\{\hat{A}_i\}$ would be equal to $\{A_i\}$. In practical cases thermal noise is present at the receiver input and it is necessary to estimate the amplitude vector $\{\hat{A}_i\}$ by solving the overdetermined equation set (4.33) using a much longer data record.

Equation (4.33) can be solved using linear algebra methods [20]. Let us assume that X_R denotes the row vector containing N received signal samples, namely, $n = 0$ to $n = N - 1$, and X_T denotes the transmitted signal vector also containing N signal samples ($n = 0$ to $n = N - 1$). Furthermore, let $X_T(l)$ be the transmitted signal vector shifted in the time domain by l samples, containing N signal samples from $n = -l$ to $n = N - 1 - l$. Let X_M be a transmitted signal matrix of dimension N I, consisting of transmitted signal vectors $X_M = \{X_T(0), X_T(1), \ldots, X_T(I-1)\}$. Let A be a column vector containing unknown clutter complex amplitudes $\{A_i\}$ for $i = 0$ to $i = I - 1$. Equation (4.29) can then be rewritten in the following matrix form:

$$X_R = X_M A \tag{4.34}$$

The unknown coefficient vector A can be estimated using the mean-square solution of (4.34), which, according to the standard method of linear algebra, can be written as:

$$\hat{A} = (X_M{}' X_M)^{-1} X_M' X_R \tag{4.35}$$

The direct use of formula (4.35) is numerically ineffective, while it is necessary to calculate the inversion of the matrix $X_M' X_M$ of $I \times I$ size. A more numerically effective method is based on solving the linear equation set in the form:

$$X'_M X_M \hat{A} = X'_M X_R \tag{4.36}$$

using, for example, the Gauss elimination method [20].

After finding the estimates of the complex amplitudes $\{\hat{A}_i\}$ it is relatively easy to remove the ground clutter echoes from the received signal. After this operation a clutter free signal should be obtained. The removal procedure can be described by the following formula:

$$x_{RC}(nT_s) = x_R(nT_s) - \sum_{i=0}^{I} \hat{A}_i x_T((n-i)T_s) \tag{4.37}$$

The received signal $x_{RC}(nT_s)$ obtained in this way (after the clutter removal procedure) can be next used for range-Doppler calculation and moving target detection. It should be pointed out that the procedure described above can effectively remove not only clutter echoes, but also the direct signal crosstalk. It results from the fact that the crosstalk signal can be also modeled by (4.32).

The procedure presented may also be derived using a linear space concept [22]. The received signal can be treated as a single point in the multidimensional signal space (see Figure 4.15). The dimension of this linear space is N (the number of received or reference signal samples in the processed data block). Usually this dimension is in the range of 10^5 to 10^7.

In this approach the clutter removal procedure is based on the concept of the decomposition of the signal space into two subspaces, namely, the clutter

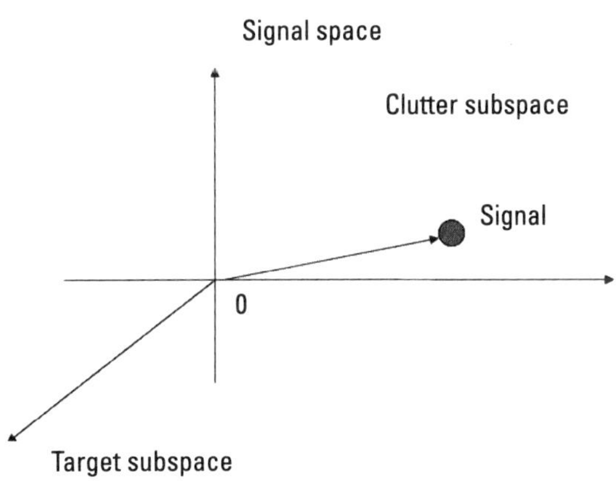

Figure 4.15 The received signal as a single point in multidimensional signal space.

and target subspace. In this situation the most straightforward method of clutter reduction will be the projection of the received signal onto the target subspace (see Figure 4.16). Such a method, however, is very difficult to implement as the dimension of the target subspace is very large.

The alternative method is based on calculation of the clutter signal and its subtraction from the received one. The dimensionality of the clutter subspace, usually in the order of 10^2 to 10^4, is usually much smaller than the dimensionality of the target subspace. For this reason it is much easier to calculate the projection of the received signal on the clutter subspace. Such a projection (see Figure 4.17) is equivalent to the estimation of the clutter signal.

In the next step the clutter signal is subtracted from the measurement signal. As a result, one obtains the clutter-free measurement signal, consisting of the receiver thermal noise and the target echoes as presented in Figure 4.18.

The methods discussed above are equivalent to each other only if the target subspace is orthogonal to the clutter subspace. The situation is completely different when there is no orthogonality between these subspaces and under the assumption that the clutter echo power is much higher than the target echo power. In such a situation the second method is much more effective and

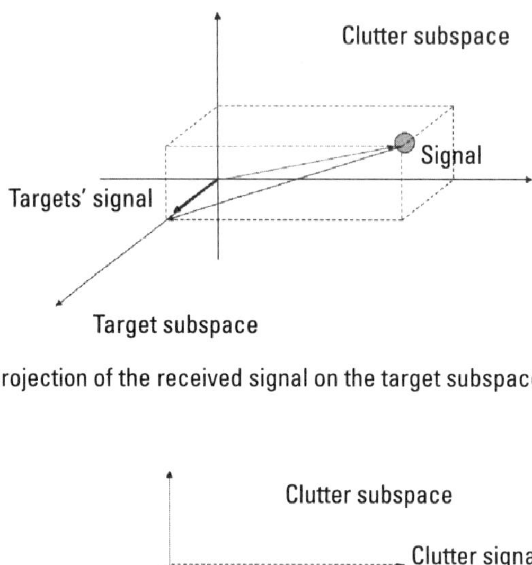

Figure 4.16 The projection of the received signal on the target subspace.

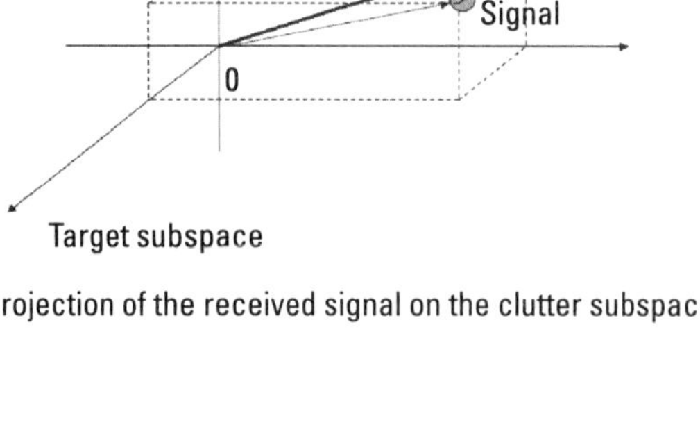

Figure 4.17 The projection of the received signal on the clutter subspace.

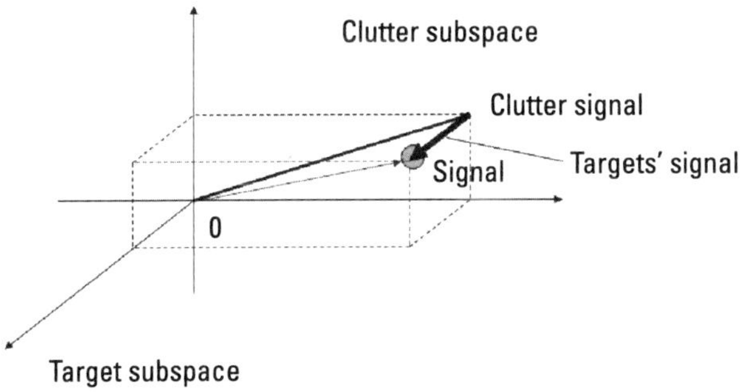

Figure 4.18 Calculation of the clutter-free target's signal.

ensures almost total clutter removal. As a result part of the target echo power will also be removed, but the loss of sensitivity is relatively small (usually below 1 dB).

The zero-Doppler clutter subspace is spanned over the vectors formed by the time-delayed copies of the reference signal, where time delay varies from zero to the delay corresponding to the maximum assumed clutter distance. The maximum clutter distance for a ground-based radar is limited by the radar horizon in the location it is placed or by the power budget; then such budget limits the maximum distance for which the clutter echoes have sufficient echo power to be detected by radar (of course after removing strong components).

For the infinitely long observation time the time-shifted versions of the noise signal are orthogonal to each other. In this case it would be possible to use the simple orthogonal projection of the signal on each clutter base vector as presented in Figure 4.19. In all practical cases observation time is finite and the vectors forming the clutter subspace do not form the orthogonal base. For that reason the more elaborate algorithms for clutter suppression have to be used for that purpose. One possibility is to solve the linear equation set described by (4.34) (see Figure 4.20). Another possibility is based on forming the orthogonal base vector set using the linear transformation of the clutter base vectors.

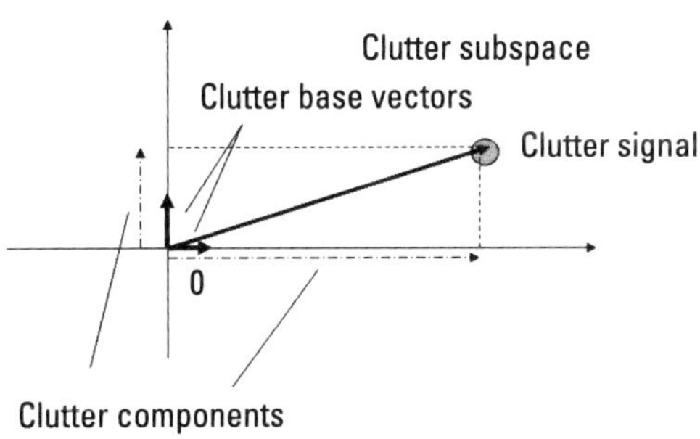

Figure 4.19 Projection of the signal on orthogonal clutter base vectors.

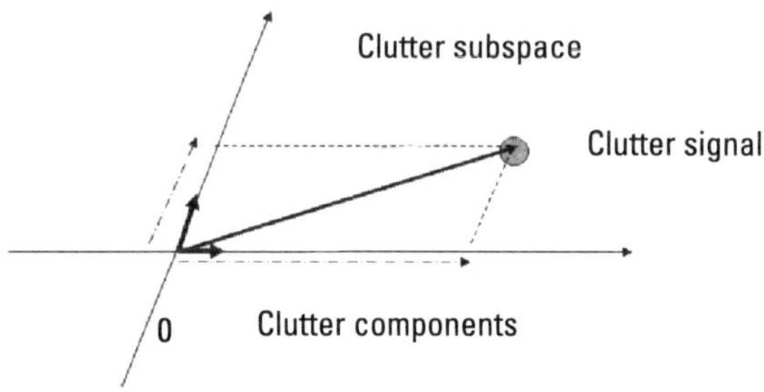

Figure 4.20 Projection of the signal on nonorthogonal clutter base vectors.

Another idea of clutter base orthogonalization is the use of Gram-Schmidt process as presented in Figure 4.21. The first base vector remains unchanged. The second orthogonal base vector is formed from the second vector by subtracting from that vector its projection on the first vector. The consecutive orthogonal base vectors are formed from the consecutive original base vectors by subtracting their projection on the already constructed orthogonal base vectors. The method described above is rather ineffective as it requires $I(I - 1)/2$ vector projections and subtractions. It can be used only for a small I number.

The signal orthogonalization can also be performed using the lattice filter [10, 21, 23, 24] presented in Figure 4.22. This method requires much smaller computational power than the previous one but only provides good orthogonalization for long data blocks.

The lattice filter presented in Figure 4.22 consists of $I - 1$ identical blocks, producing the backward ($X_{b(i)}$) and forward ($X_{f(i)}$) prediction error vectors on the basis of the backward and forward prediction error vectors from the previ-

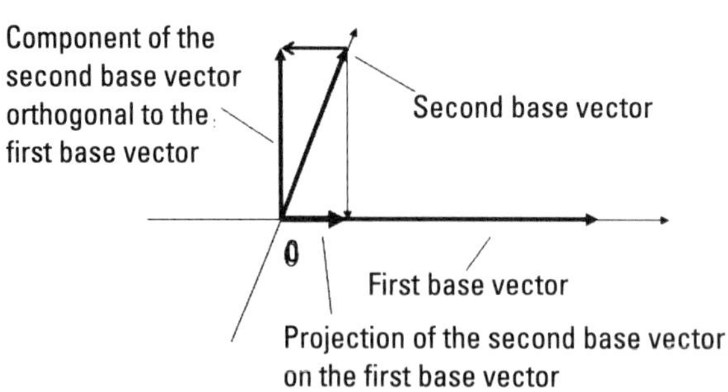

Figure 4.21 Orthogonalization of the clutter base vectors.

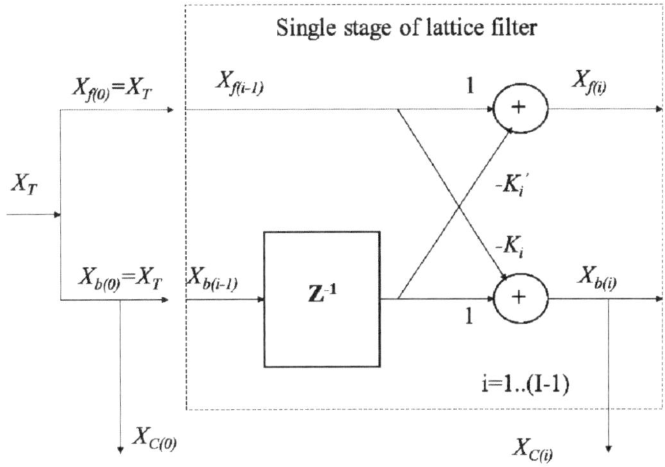

Figure 4.22 The lattice filter-based clutter subspace orthogonalization.

ous stage ($X_{b(i-1)}$ and $X_{b(t-1)}$). At both inputs of the first stage the transmitted (reference) signal vector X_T is applied.

The lattice filter backward prediction errors from consecutive filter stages are orthogonal to each other, and thus the lattice structure can be used for clutter base orthogonalization. The partial correlation coefficients for each stage are calculated using the following formula:

$$K_i = \frac{2\sum_{n=i+1}^{N} x_{b(i-1)}(n-1)x'_{f(i-1)}(n)}{\sum_{n=i+1}^{N}\left(\left|x_{b(i-1)}(n-1)\right|^2 + \left|x_{f(i-1)}(n)\right|^2\right)} \tag{4.38}$$

The forward prediction error signal $x_{f(i)}$ is calculated using the following formula [10]:

$$x_{f(i)}(n) = x_{f(i-1)}(n) - K_i' x_{b(i-1)}(n-1) \tag{4.39}$$

and the backward prediction error signal $x_{b(i)}$ is calculated using the formula:

$$x_{b(i)}(n) = x_{b(i-1)}(n-1) - K_i' x_{f(i-1)}(n) \tag{4.40}$$

The backward prediction error signals form the orthogonal clutter base vectors:

$$x_{C(i)}(n) = x_{b(i)}(n), i = 0...I-1 \tag{4.41}$$

Those orthogonal base vectors are used further to cancel the clutter from the received signal. The cancellation process, presented in Figure 4.23, can be described by the formula:

$$x_{R(i+1)}(n) = x_{R(i)}(n) - h_i' x_{c(i)}(n) \tag{4.42}$$

where $x_R(0) = x_R$ is the received signal, $x_R(i)$ is the signal after the ith stage of clutter cancellation, and h_i' is the correlation coefficient calculated using the formula:

$$h_i = \frac{\sum_{n=0}^{N-1} x'_{R(i)}(n) x_{c(i)}(n)}{\sum_{n=0}^{N-1} \left| x_{c(i)}(n) \right|^2} \tag{4.43}$$

In Figures 4.24 through 4.29 the simulated result of the ground clutter removal procedure is presented. The received signal consists of the direct signal (+40 dB over receiver noise), several ground clutter echoes (0.. 20 dB over receiver noise), a strong moving target ccho (S/N = 0 dB) at a range of 15 km, and a weak echo moving target echo (S/N = –16 dB) at a range of 27 km. The result of the correlation processing of the received signal is presented in Figure 4.24. Only the direct signal and the strong ground clutter are visible. The processing noise floor is at the level of 95 dB and it masks moving-target echoes. The result of the classical MTD procedure, which filters out the zero-Doppler

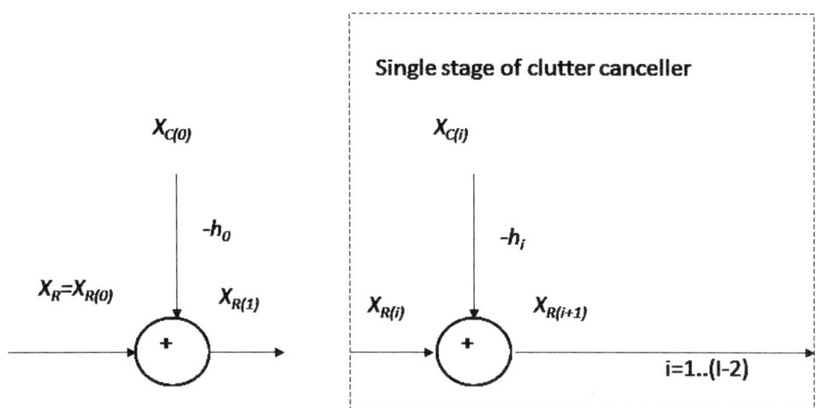

Figure 4.23 Clutter cancellation based on lattice filter clutter base orthogonalization.

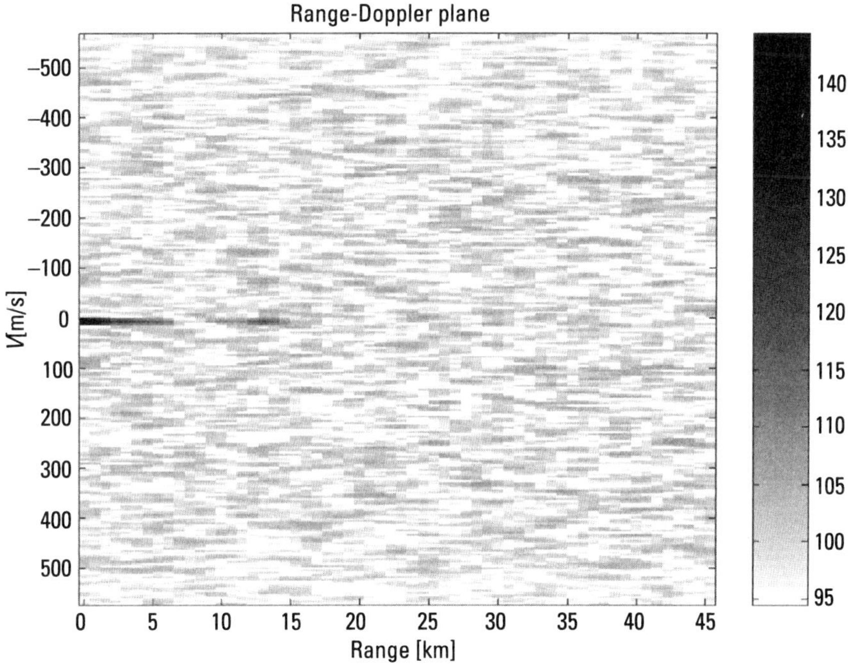

Figure 4.24 The magnitude of the range-Doppler cross-correlation function for a measurement signal containing a direct signal and ground clutter. Grayscale is in decibels.

components, is presented in Figure 4.25. The direct signal, as well as the ground clutter, is filtered out, but the processing noise floor is at the same level. Both moving targets are masked by the sidelobes of the direct signal and the ground clutter echoes. The removal of the direct signal improves the situation. The result of the range-Doppler processing of the received signal after the direct signal removal procedure is presented in Figure 4.26. The processing noise floor is now reduced by 15 dB to the level of 80 dB, and the strong moving target echo is now clearly visible together with the ground clutter echoes. The weak target echo is still below the processing noise floor (sidelobes of ground clutter). Again, the usage of the classical MTD technique, presented in Figure 4.27, removes the ground clutter, but does not remove the ground clutter sidelobes, and thus the weak target is still hidden in the ground clutter sidelobes. The procedure for ground clutter removal described above completely eliminates the direct signal and the ground clutter echoes together with their sidelobes, as depicted in Figure 4.28. In the case presented, the processing noise floor has been reduced by 35 dB to the level of 60 dB. Now both moving target echoes are clearly visible.

Just for comparison, in Figure 4.29 we present the result of the range-Doppler processing of the received signal in the case where no direct signal

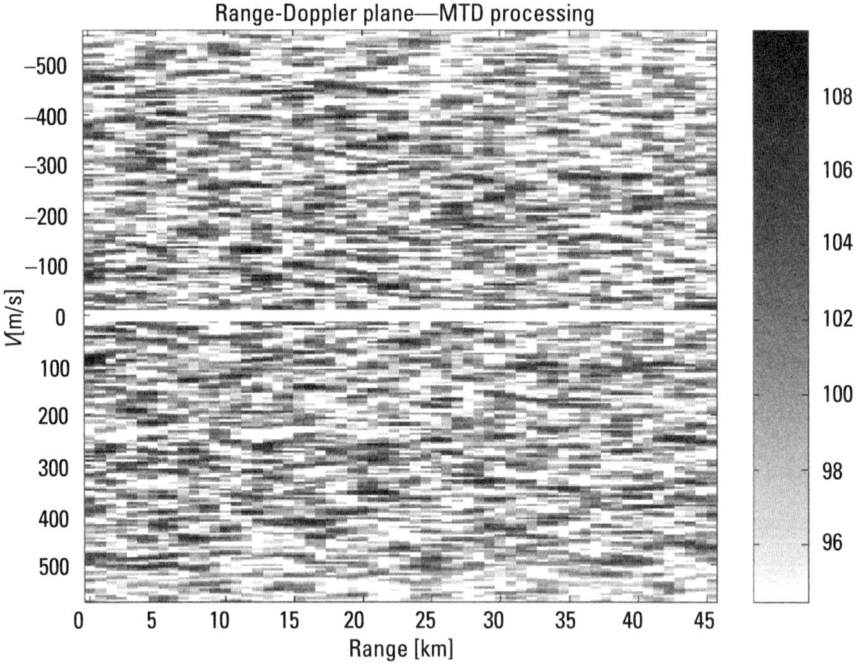

Figure 4.25 The magnitude of the range-Doppler cross-correlation function for a measurement signal containing a direct signal and ground clutter after MTD processing. Grayscale is in decibels.

or ground clutter was present. Those two figures (Figures 4.28 and 4.29) are almost identical (it is possible to notice only slight differences in the processing sidelobes structure) and it shows the effectiveness of the described procedure for the ground clutter echo removal.

The block adaptive method presented in this chapter is only one of several possible methods used for ground clutter cancellation in noise radar. Almost all adaptive filtration procedures can be applied for this purpose [4, 25–27]. However, block algorithms perform better when block processing is applied. Recursive algorithms are better suited for continuous signal reception and block partitioning after a ground clutter removal procedure. Recursive algorithms usually require a bit more computational power, but do not have to remember whole blocks of data so they are better suited for fast signal processing, especially using field programmable gate arrays (FPGAs).

4.3.3 Doppler-Spread Clutter Cancellation

The method presented in Section 4.2.2 is very efficient for the direct signal and for the zero-Doppler ground clutter cancellation. As it was shown in the

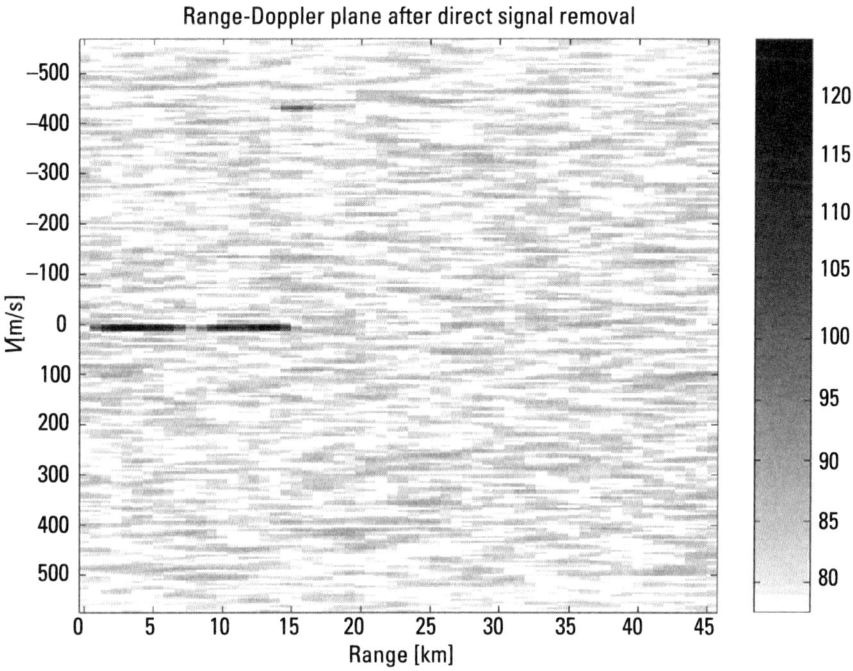

Figure 4.26 The magnitude of the range-Doppler cross-correlation after the direct signal removal procedure. Grayscale is in decibels.

presented example, the described method can completely cancel both the direct signal and the ground clutter with their processing sidelobes.

In many real radar scenarios the clutter model is much more complicated. The Doppler shift of the clutter echo is not always zero, while the clutter scatterers may be swung or moved by the wind, vibrate, or have a more complicated motion. As stated in [28], the ground clutter spectrum very often has a Gaussian shape and Doppler spread equivalent to the clutter movement with a speed from 0.01 to 5 m/s. The Doppler spread is different for different areas (ground types). In urban areas buildings can be swung by the wind [28–30]. The tops of high buildings may have velocities up to 0.1 m/s. In urban areas, except for the reflections from buildings there are a lot of reflections from vehicles and pedestrians. The velocity spread for pedestrians is in the range of 2 m/s, while for vehicles it can reach 30 m/s. Depending on the application, the pedestrians and vehicles can be treated as the targets of interest or as clutter. For vegetation areas (fields, forests, meadows), leaves and branches can be swung by the wind with the velocity spread of almost 1 m/s. On the sea's surface there are many types of waves that can travel in various directions with different velocities. Sea clutter usually has a velocity spread in the range of several m/s and also a nonzero mean Doppler frequency. Much more complicated is weather clutter. In

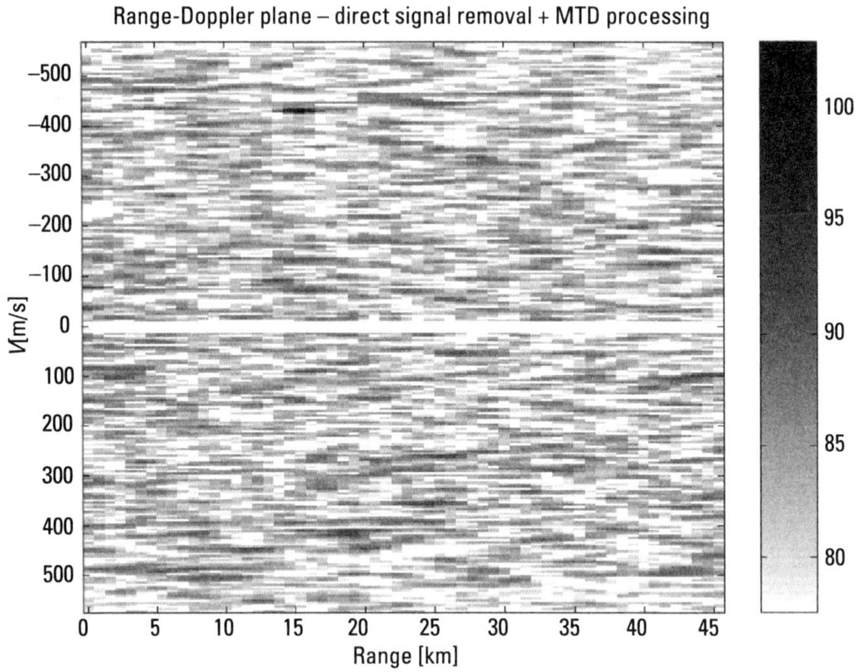

Figure 4.27 The magnitude of the range-Doppler cross-correlation function after the direct signal removal procedure and MTD processing. Grayscale is in decibels.

heavy wind conditions the mean clutter velocity can reach tens of m/s, with a very wide velocity spectrum caused by wind turbulence (especially during storm conditions).

To detect the weak targets in the presence of Doppler spread clutter, it is necessary to attenuate those clutter echoes [31, 32]. The noise radar Doppler-spread clutter echo signal can be modeled by the weighted sum of the time delayed and Doppler shifted copies of the transmitted signal, where the maximum time delay is limited by the maximum range of the clutter, and the maximum Doppler shift is limited by the maximum clutter velocity. For the sake of simplicity it is assumed that there is the Doppler separation between the clutter and the targets. The clutter signal can then be modeled by the discrete formula [32]:

$$x_R(nT_s) = \sum_{i=0}^{I} \sum_{k=-K}^{K} A_{i,k} x_T\left((n-i)T_s\right) \exp\left(j2\pi kn/N\right) \tag{4.44}$$

This formula describes the zero-mean Doppler clutter, spread from a normalized frequency by $-K/N$ to K/N. The above formula also incorporates

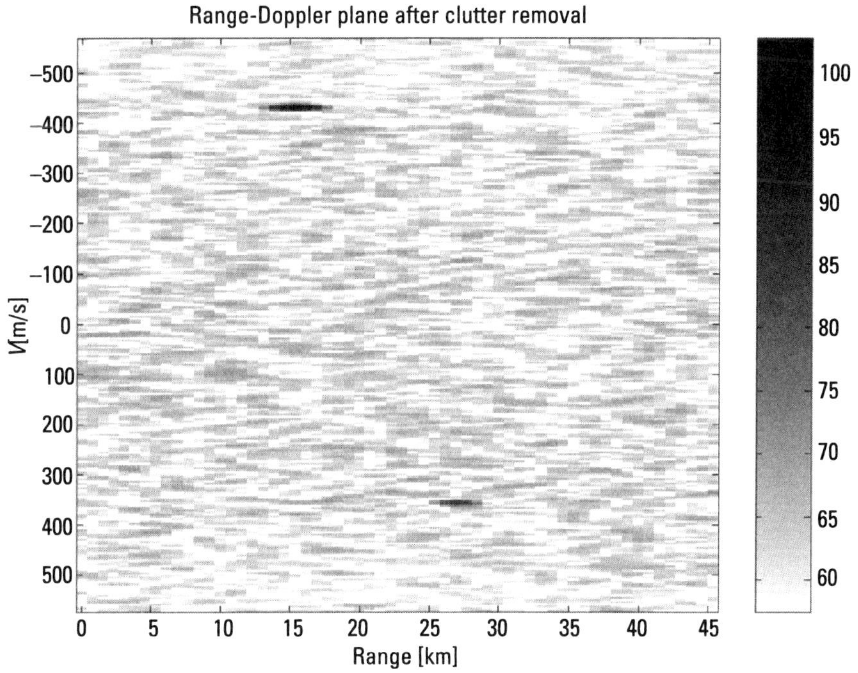

Figure 4.28 The magnitude of the range-Doppler cross-correlation function after the direct signal and ground clutter removal procedure. Grayscale is in decibels.

the direct signal and the ground clutter echoes. For the case of nonzero mean Doppler, the model can be modified slightly:

$$x_R(nT_s) = \sum_{i=0}^{I} \sum_{k=K_{mni}}^{K_{\max}} A_{i,k} x_T\left((n-i)T_s\right)\exp\left(j2\pi\, kn/N\right) \tag{4.45}$$

and in this case the clutter Doppler frequency is in the range from $K_{\min}/N$ to $K_{\max}/N$. It is also possible to modify the Doppler spread with the range that leads to the formula:

$$x_R(nT_s) = \sum_{i=0}^{I} \sum_{k=K_{mni}(i)}^{K_{\max}(i)} A_{i,k} x_T\left((n-i)T_s\right)\exp\left(j2\pi\, kn/N\right) \tag{4.46}$$

Under further consideration, the first model described by (4.44) will be applied. All presented results however can be easily adapted to more elaborate clutter models, described by (4.39) and (4.40). It is worth noting that those clutter models can be applied both to medium integration and long integration

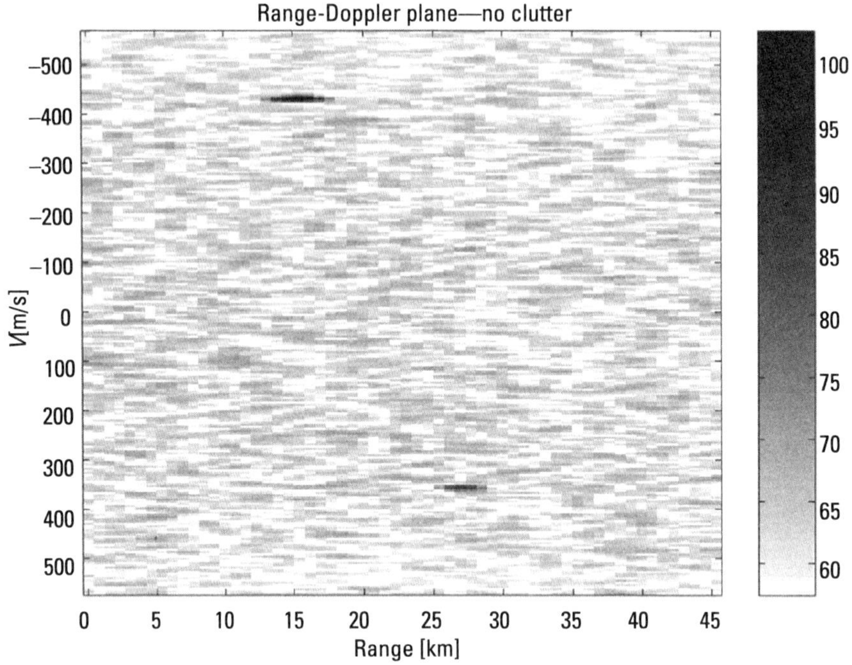

Figure 4.29 The magnitude of the range-Doppler cross-correlation function: simulation without direct signal and ground clutter. Grayscale is in decibels.

time noise radars if the maximum clutter velocity is much lower than described by the constraint (2.15).

As in the previous case, clutter cancellation can be performed by estimating the unknown complex amplitudes A_{ik} of the clutter echoes and subtracting the modeled clutter signal from the received one, obtaining the clutter-free signal x_{RC} according to the following formula:

$$x_{RC}(nT_s) = x_R(nT_s) - \sum_{i=0}^{I} \sum_{k=-K}^{K} \hat{A}_{i,k} x_T\left((n-i)T_s\right) \exp\left(j2\pi kn/N\right) \quad (4.47)$$

The estimation of the unknown vector of complex amplitudes $\{A_{i,k}\}$ of clutter echoes can be calculated by solving an overdetermined equation set in the form:

$$\left\{ x_R(nT_s) = \sum_{i=0}^{I} \sum_{k=-K}^{K} \hat{A}_{i,k} x_T\left((n-i)T_s\right) \exp\left(j2\pi kn/N\right) \right\} \quad (4.48)$$

As in the case of zero-Doppler ground clutter, it is possible to construct the received signal vector X_R as the row vector containing N received signal samples from $n = 0$ to $n = N - 1$. The transmitted signal matrix X_M (dimension $NxIx(2K + 1)$) now consists of time and Doppler shifted copies of the transmitted signal. Let us denote that $X_T(l,k)$ is the transmitted signal vector shifted in time by l samples, multiplied element by element by frequency shift vector $\{\exp(-2\pi nk/N\}$. The vector $X_T(l,k)$ can then be written in the form:

$$X_T(l,k) = \left\{x_T(n-l)\exp(-j2\pi nk/N\right\}_n \tag{4.49}$$

Let us now construct the transmitted signal submatrix $X_M(k)$ with Doppler shift k/N. The submatrix is in the form:

$$X_M(k) = \left\{X_T(0,k), X_T(1,k), \ldots, X_T(I-1,k)\right\} \tag{4.50}$$

The transmitted signal matrix X_M is now constructed using the $X_M(k)$ submatrix, and has the form:

$$X_M = \left\{X_M(-K), X_M(-K+1), \ldots, X_M(K)\right\} \tag{4.51}$$

By analogy, let us denote $A(k)$ to be an unknown amplitude column vector for Doppler shift k/N:

$$A(k) = \left\{A_{0,k}; A_{1,k}; \ldots; A_{I-1,k}\right\} \tag{4.52}$$

where ; denotes a column vector constructor.

The unknown amplitude vector A is a column vector constructed using $A(k)$ subvectors and has the form:

$$A = \left\{A(-K); A(-K+1); \ldots; A(K)\right\} \tag{4.53}$$

Equation (4.48) can be then rewritten in the matrix form (4.34), and the unknown coefficient vector A can be estimated using the mean-square solution of (4.34), which is described by (4.35).

However, the direct usage of the method described above is very limited. Such a method was not numerically effective in the case of ground clutter, where the dimensionality of clutter subspace was in the order of 10^2 to 10^4 and now the dimensionality of clutter subspace is much higher, in the order of 10^3 to 10^5. Except for a high numerical complexity of the method, we can expect relatively high numerical errors during matrix inversion and thus limited quality of the clutter suppression.

Instead of solving the overdetermined equation set, it is again possible to apply the concept of the projection of the received signal on the clutter subspace. The clutter subspace is now spanned on time-delayed and Doppler-shifted copies of the transmitted signal. The matrix X_M (4.51) contains all clutter base vectors. Again, as in the case of the ground clutter removal, the clutter subspace base vectors do not form the orthogonal base. The orthogonalization of clutter subspace is now much more difficult. The classical orthogonalization procedure described in Section 4.2.3 has very high computational complexity, so suboptimal methods are required.

The presented method is based on the observation that in most cases the dominant part of clutter energy originates from zero-Doppler or almost zero-Doppler echoes. The use of the lattice filter can produce a zero-Doppler orthogonal base in the form of $\{X_{C(i)}\}$ (4.41). Applying the frequency shift to those vectors, one can obtain an almost orthogonal clutter base, constructed from vector $X_{C(i,k)}$:

$$X_{C(i,k)} = \left\{x_{C(i)}(n)\exp(j2\pi nk/N)\right\}_n \tag{4.54}$$

The idea of simple clutter removal using a nonorthogonal base is presented in Figures 4.30 and 4.31. The signal at the input of the consecutive stage is projected on the consecutive base vector. The projected clutter component is subtracted from the signal, and the procedure is continued for the next base vector. In Figure 4.30 the removal of the first clutter component is presented. The removal of the second component is presented in Figure 4.31. After finishing the procedure the clutter is not removed completely because of a lack of orthogonally of the base vectors; some residual clutter remains in the signal. This clutter can be further reduced by applying the clutter removal procedure again. When the clutter base vectors are already created, the computational cost

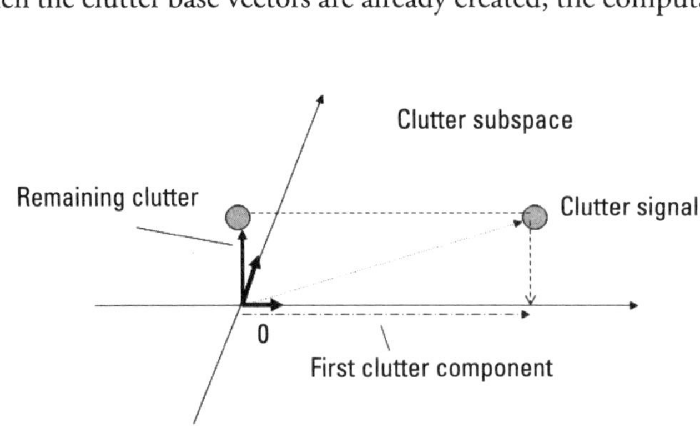

Figure 4.30 Removal of the clutter component related to the first clutter base vector.

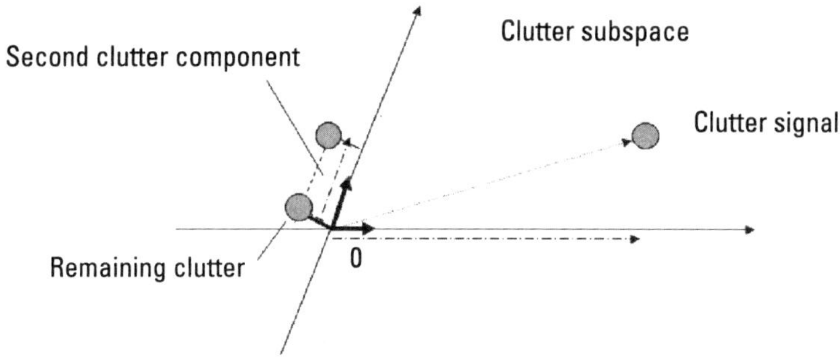

Figure 4.31 Removal of the clutter component related to the second clutter base vector.

of the repetition of the clutter removal procedure is moderate, but it is necessary to store all clutter base vectors in data memory, which can be difficult in practical implementation.

The Doppler-spread clutter cancellation process, the block diagram of which is presented in Figure 4.32, can be described by:

$$
\begin{aligned}
x_{R(i,k+1)}(n) &= x_{R(i,k)}(n) - h'_{i,k} x_{c(i,k)}(n) \\
x_{R(i,-K)}(n) &= x_{R(i-1,K+1)}(n)
\end{aligned}
\tag{4.55}
$$

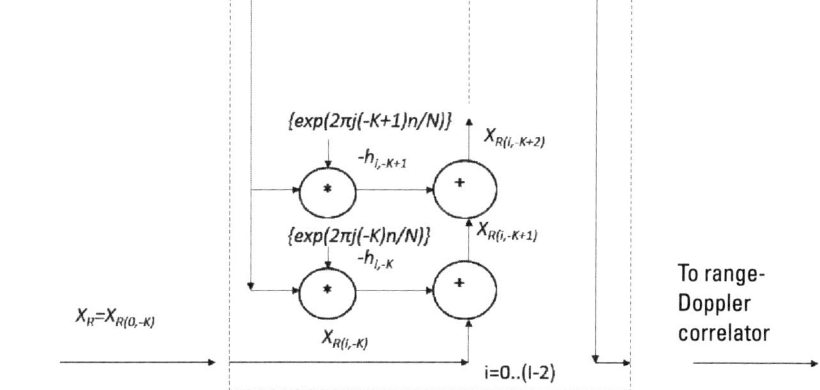

Figure 4.32 The structure of the Doppler spread clutter canceler.

The correlation coefficient $h'_{i,k}$ can be calculated using the following formula:

$$h_{i,k} = \frac{\sum_{n=0}^{N-1} x'_{R(i,k)}(n)\, x_{c(i,k)}(n)}{\sum_{n=0}^{N-1} \left| x_{c(i,k)}(n) \right|^2} \tag{4.56}$$

The quality of the single-pass Doppler spread cancellation depends on the angle between the clutter base vectors. Instead of calculating the angle between vectors, it is possible to calculate the base vector product. For orthogonal vectors such a coefficient is equal to zero, and for the proposed base vector it is equal to:

$$\begin{aligned} X_{C(i_1,k_1)} \otimes X_{C(i_2,k_2)} &= \sum_{n=0}^{N-1} x'_{C(i_1,k_1)}(n)\, x_{C(i_2,k_2)}(n) \\ &= \sum_{n=0}^{N-1} x'_{C(i_1)}(n)\, x_{C(i_2)}(n) \exp\left(j2\pi n(k_2 - k_1)/N\right) \end{aligned} \tag{4.57}$$

Analyzing the above formula, it is easy to notice that for each selected frequency the different time-delayed base vectors are orthogonal to each other:

$$X_{C(i_1,k_1)} \otimes X_{C(i_2,k1)} \underset{i1 \neq i2}{=} \sum_{n=0}^{N-1} x'_{C(i_1)}(n)\, x_{C(i_2)}(n) = 0 \tag{4.58}$$

For the selected delay the vector product is equal to the Fourier transform of the square of the absolute value (amplitude) of the $X_{C(i)}$ vector for the frequency equal to the frequency shift between vectors:

$$X_{C(i_1,k_1)} \otimes X_{C(i_1,k_2)} = \sum_{n=0}^{N-1} \left| x'_{C(i_1)}(n) \right|^2 \exp\left(j2\pi n(k_2 - k_1)/N\right) \tag{4.59}$$

In general cases the vector product for $k_2 \neq k_1$ is more than $Bt_i \approx N$ smaller than it is for $k_2 = k_1$, so the clutter cancellation procedure should work pretty well. For constant amplitude noise (constant amplitude base vectors) this product is equal to zero for $k_2 \neq k_1$, and this is the reason why constant amplitude noise is preferred for a noise radar working in an environment with Doppler spread clutter. For the cross-terms, for $k_2 \neq k_1$ and $i_2 \neq i_1$, the vector product is equal to the Fourier transform of the product of the vectors $X_{C(i_1)}$ and $X_{C(i_1)}$

for the frequency $(k_2 - k_1)/N$. While both vectors consist of independent white noises, their product should have a white spectrum, and the mean value of the product (4.57) should also be smaller than the mean vector energy by factor $Bt_i \approx N$.

In Figures 4.33 through 4.37 a simulated result of the Doppler spread clutter removal procedure is presented. The scenario consists of one point target echo (at the range of 7 km and a velocity of −22 m/s) and 100 clutter point echoes, distributed randomly in range and Doppler. The Gaussian distribution for the Doppler spread has been chosen. The result of the correlation processing of the received signal is presented in Figure 4.33. Only the strong zero-Doppler clutter is visible. The clutter returns reach a level of 74 dB. The processing noise floor is at the level of 25 dB and it masks the moving target echo.

The effect of the zero-Doppler removal procedure is presented in Figure 4.34. After removing the zero-Doppler clutter, it is possible to observe the Doppler spread clutter, which reached the level of 57 dB. The maximum clutter level was reduced by 18 dB, but the processing noise floor was reduced only by 13 dB to the 12-dB level. Reduction of the noise floor is smaller than the reduction of the maximum clutter level because now there are many contributors

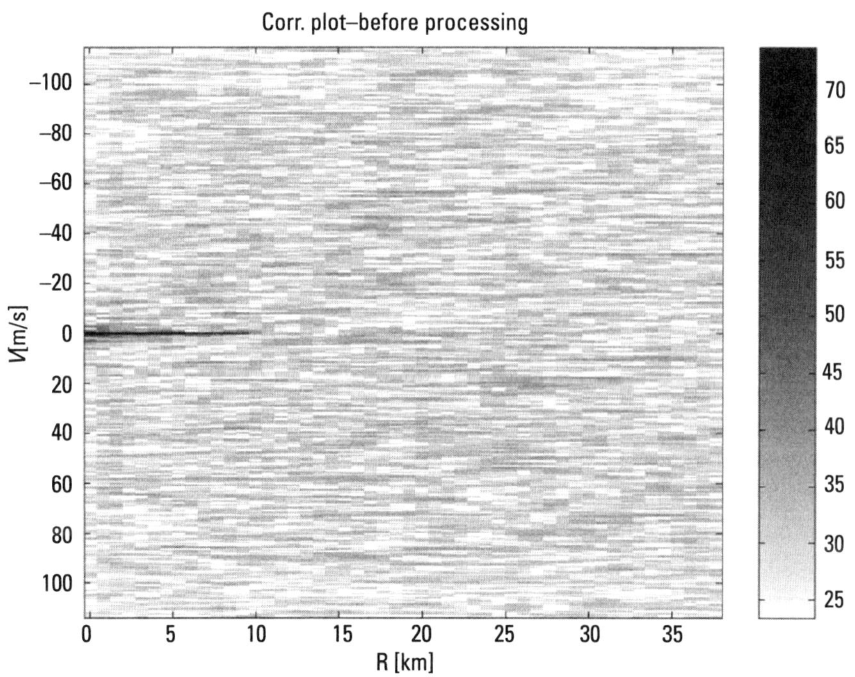

Figure 4.33 The magnitude of the range-Doppler cross-correlation function for a measurement signal containing Doppler spread clutter. The zero-Doppler clutter is visible, and the target echo is hidden in clutter sidelobes. Grayscale is in decibels.

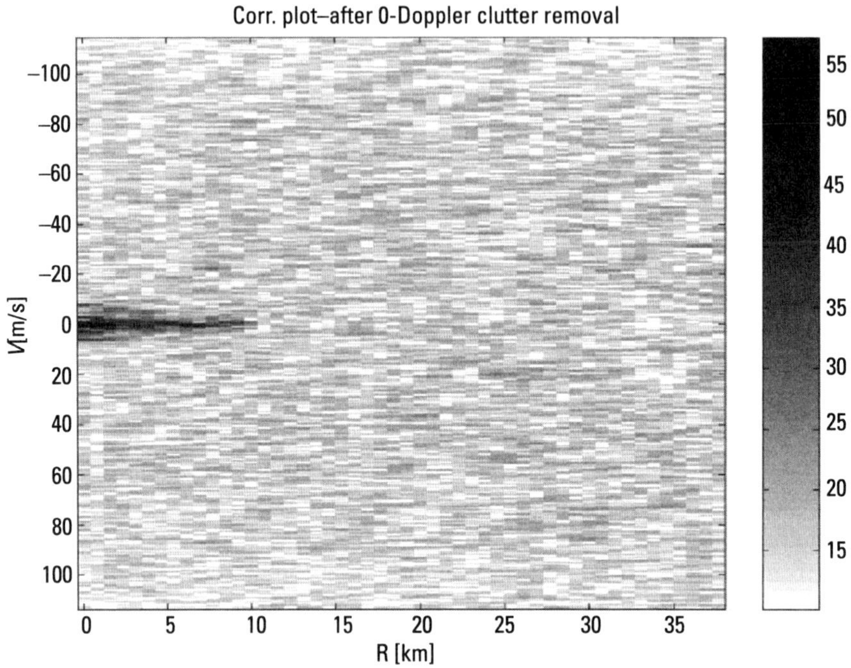

Figure 4.34 The magnitude of the range-Doppler cross-correlation function after the zero-Doppler clutter removal procedure. The Doppler spread clutter is clearly visible, and the target echo is almost invisible. Grayscale is in decibels.

to the overall level of the noise floor. Now it is possible to guess the location of the target echo, but the target echo level is still at the same level as the higher processing sidelobes. Using classical automatic detection procedures [28] based on a constant false alarm rate detector, it would be very difficult to detect the target. After the application of the Doppler spread clutter removal procedure described by (4.56), the processing noise floor was reduced by a further 20 dB reaching the level of −8 dB. The target echo is now clearly visible as depicted in Figure 4.35, but the residual clutter remains in the clutter rejection region (range from 0 to 12 km, velocities from −10 m/s to 10 m/s). The second use of the Doppler spread clutter rejection procedure (see Figure 4.26) significantly clears the clutter rejection area and decreases the processing noise floor by an additional 3 dB, to the level of −11 dB. In this example the clutter cancellation procedure increased the dynamic range of the radar by 36 dB, extending eight times the radar detection range.

Just for comparison, in Figure 4.37 the result of the Doppler spread clutter reduction based on the overdetermined equation set solving is presented. The result is very similar to the case of the double usage of the previous algorithm (compare it with Figure 4.36). The difference in the processing noise

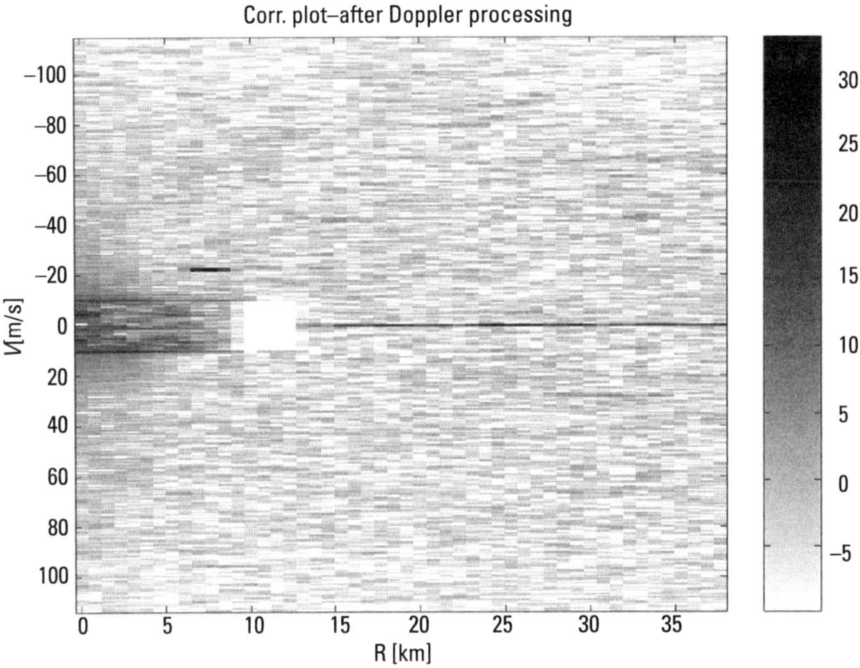

Figure 4.35 The magnitude of the range-Doppler cross-correlation function after Doppler-spread cancellation. The target echo is clearly visible, and the clutter residue is present in the clutter rejection area. Grayscale is in decibels.

floor is within 0.5 dB, but in this case the clutter rejection region has been cleared completely.

4.4 Robustification of the Noise Radar Signal Processing

Target detection of the target of interest in a noise radar is based on the correlation of the transmitted signal with the received one, which is equivalent to the matched filtering of the received signal. Matched filtering is optimal (in the mean square sense) only when one target is present in the received signal and the receiver noise (thermal noise of the receiver, external noise, interference) has a Gaussian probability density function (PDF). In the previous chapters analyses of the multitarget problem were presented and a suboptimal solution based on the CLEAN concept was proposed. Now we will discuss the problem of the occurrence of non-Gaussian interferences at the input of the correlation receiver.

In many practical cases the input noise consists of impulse interferences and the Gaussian part originates from the receiver thermal noise. As a result, the receiver noise has a non-Gaussian PDF with heavy tails and the correlation

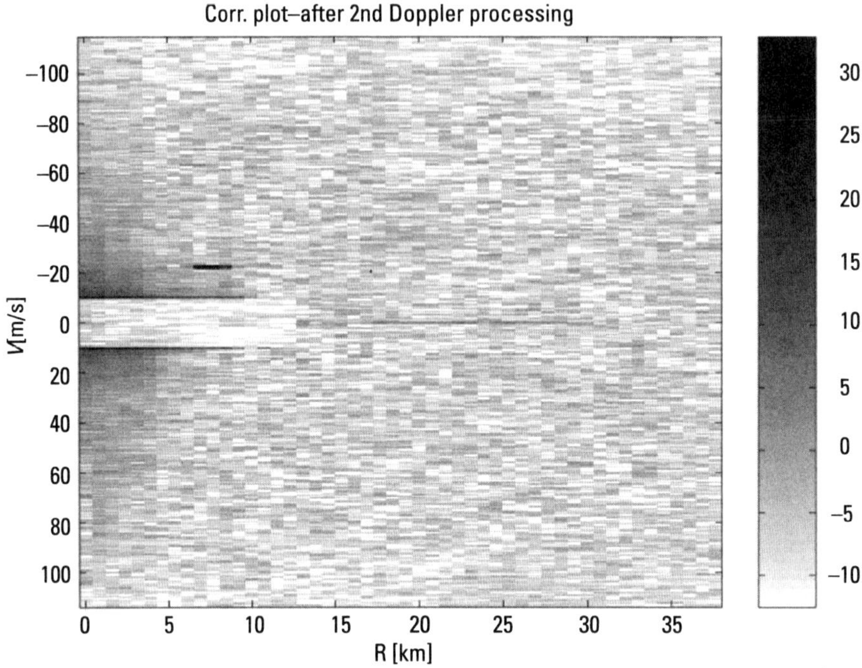

Figure 4.36 The magnitude of the range-Doppler cross-correlation function after the second Doppler-spread cancellation. The target echo is clearly visible, and the clutter residue has been removed almost completely. Grayscale is in decibels.

process is no longer optimal. To improve noise radar sensitivity in the case of a non-Gaussian PDF of the input noise, radar engineers use nonlinear processing, which is usually known as robustification of the radar detection. The name comes from the fact that after such a procedure, a radar is much less sensitive (so it is robust) to the changes of the probability density function of input noises and interferences. Just to show the existence of the problem, the real part of the signal recorded during a noise radar experiment at the Warsaw University of Technology is presented in Figure 4.38. The measurement was performed from the roof of the university situated in the center of Warsaw, in a typical urban environment. It is easy to see the impulsive character of the signal. In Figure 4.39 the probability density function of the real part of the received signal is presented. For reference normal distribution was plotted in a dashed line. In the one sigma region there is a pretty good agreement between the measured and Gaussian PDF, but at further regions (for higher amplitudes) the heavy tails are clearly visible.

Next a robustification procedure will be shown in a noise radar application, and the effective solution of detector robustification in case of clutter will be shown.

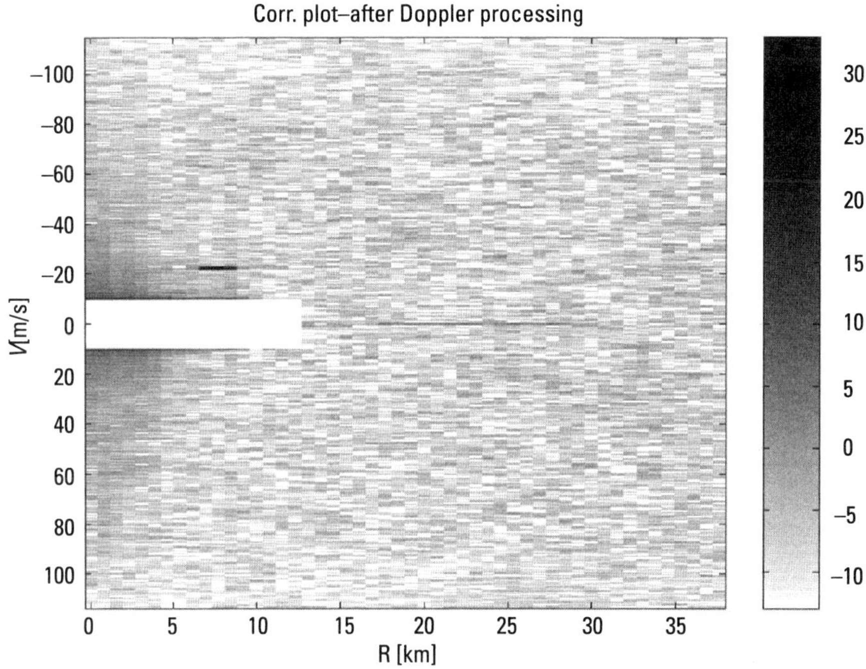

Figure 4.37 The magnitude of the range-Doppler cross-correlation function after Doppler spread clutter removal based on the matrix equation solution. The clutter is removed completely the from clutter rejection area. Grayscale is in decibels.

For decades, the Gaussian probability distribution has been one of the most important tools of the description of random signals in communications and signal processing theory and practice. Many methods and algorithms for signal processing have been derived on the basis of the Gaussian character of signals under consideration. One of the most important reasons for the role of the Gaussian distribution in such signal processing lies in the central limit theorem (CLT) which states that a random variable being a sum of infinitely many independent random variables with finite variances are normally (Gaussian) distributed. Thus, if there is no a priori information about the statistical model of the noise, it is reasonable to assume that the noise in an additive channel will obey the Gaussian model. This model is also very popular when there is a closed-form solution for problems where such noise exists. In the derivation of various signal processing algorithms, the Gaussian assumption leads to linear solutions, which are elegant from a mathematical point of view and usually relatively cheap in implementation. Another great advantage of the Gaussian model is its invariance under linear transformations. This means that all filtration algorithms, correlations, and Fourier transforms do not change the PDF if the input PDF is Gaussian. Moreover, if the input PDF is non-Gaussian, such a linear operation will almost turn it into Gaussian, based on

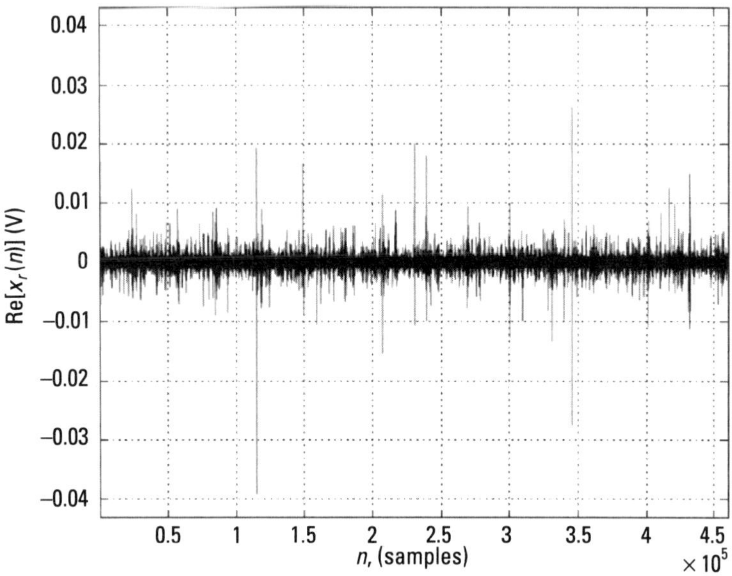

Figure 4.38 Real part of the signal recorded in the noise radar experiment. The impulsive interfereces are clearly visible.

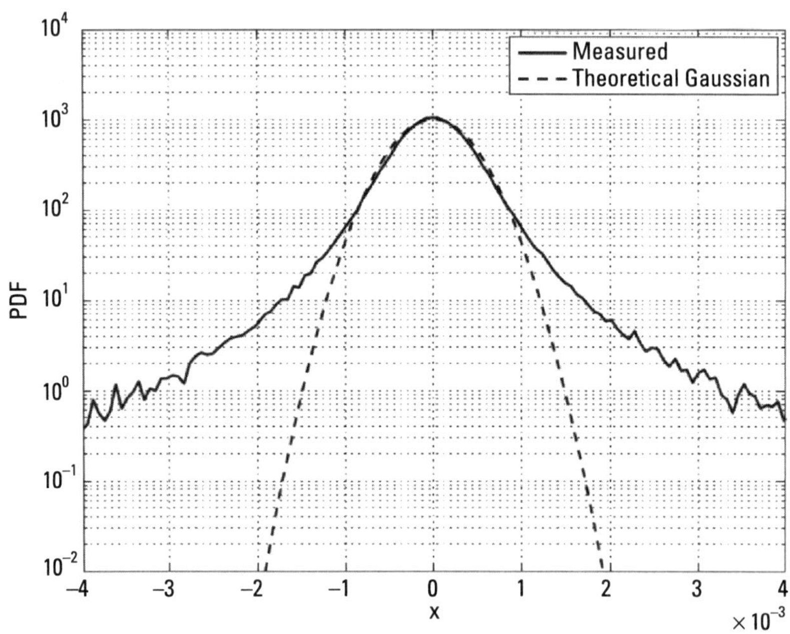

Figure 4.39 Measured probability density function of the real part of the received signal. Heavy tails are clearly visible.

the CLT. The Gaussian process is described completely by its second-order moments (the mean value and variance) and no numerically burdensome higher order statistics are necessary in this case. Moreover, we can use very attractive mathematical tools such as the Hilbert space formalism in the analysis of the Gaussian random variable.

Despite the strong theoretical motivation and practical usefulness of the Gaussian model, there are many cases where noise generation is not based on the accumulative process and in which the model is unrealistic and inadequate [33–35]. These problems are related to the impulsive nature of many signals and noises present in the physical world. The impulsive behavior manifests itself in large amplitude outliers occurring in the observed data, with a probability that is significantly higher than that predicted by the Gaussian distribution. Thus, non-Gaussian distributions with much heavier tails are required for the proper probabilistic description of signals with impulsive noises. For decades, many efforts have been made in literature to develop statistical non-Gaussian models for impulsive noises. Classical statistical-physical models of interference in radio channels were proposed by Middleton [36–38]. He introduced three different classes of electromagnetic interferences, denoted as A, B, and C. Middleton's models have been widely accepted since they describe real impulsive interferences very well. However, the mathematical complexity is a big disadvantage of these models. This is particularly true for the class B model that contains six parameters, including one which is purely empirical. Alternatives to his model are the symmetric α-stable (SαS) distributions presented in [39, 40]. The α-stable distributions constitute a different class of distributions but they can also successfully approximate Middleton's class B distributions [41]. Such distribution is difficult to handle mainly due to the lack of a closed-form expression of PDF. The another candidate which approximates Middleton's class A distribution is the ε-contaminated model, which is the mixture of two weighted Gaussian PDFs described by the following class of distributions:

$$\mathsf{F}_\varepsilon \doteq \left\{ F : F = (1-\varepsilon) N\left(0,\sigma^2\right) + \varepsilon N\left(0,\kappa\sigma^2\right) \right\} \tag{4.60}$$

where $N(0, \sigma^2)$ is the Gaussian distribution with the variance σ^2, $0< \varepsilon <1$ is a distribution parameter describing how often heavy tails occur and $\kappa >> 1$. It is possible to expand this model for a more element weighted sum of Gaussian PDFs [42].

The intuitive interpretation of the ε-contaminated model (4.60) is very simple. Most of the observed input noise is assumed to be Gaussian, but a fraction ε of observations is affected by another, heavy tails noise, which also has Gaussian distribution but with a much larger variance. Such outliers can be caused by many phenomena: atmospheric discharges, man-made interferences, communication disruptive signals (like WiFi), and others.

It is well known that almost any algorithm developed on the basis of the Gaussian assumption loses its optimal properties while the noise probability density function is not Gaussian. Even a small deviation from Gaussianity usually leads to the loss of optimality and can cause significant performance degradation of the algorithm. It is especially true in the detection process, where heavy tails can either raise the probability of a false alarm or, after proper use of the constant false alarm rate detector (CFAR), increase the detection threshold and then significantly decrease the probability of the detection of useful targets.

As was stated before, the probability density function of the noise at the output of the correlator is Gaussian almost regardless of the input probability density function, but heavy tails can change the noise floor of the processing and as the result decreases the sensitivity of the radar. This drawback can be omitted by introducing some nonlinear data transformation before the correlation in the place where the non-Gaussian probability density function exists. In many statistical problems, when the PDF of the input signal is known, this nonlinear transformation can be chosen optimally. Huber formulated in [43, 44] the minimax rules of the optimal choice of these nonlinear functions. He considered the problem of the robust estimation of the location parameter θ in the case of an independent, identically distributed observation of the form $x_i = \theta + u_i$ where u_i are described by a distribution belonging to the class (4.60). The proposed robust *M*-estimator $\hat{\theta}$ is given by the solution of a nonlinear equation:

$$\sum_{i=1}^{n} \psi\left(x_i - \theta\right) = 0 \tag{4.61}$$

where x_i represents the ith observation and ψ is the robustifying function. When the probability density function of an ε-contaminated noise is known, that is, when the parameters ε and κ are fixed and known, the choice $\psi(x) = -f'(x)/f(x)$ leads to the well-known maximum likelihood estimator of the location. Another particular choice of function ψ, namely, $\psi(\mathrm{x}) = x$, leads to the standard nonrobust estimation, optimal in the least squares (LS) sense. In this case, the solution of the estimation equation (4.61) leads to the sample mean. As it is well known, the performance of the sample mean as an estimator of location degrades substantially in the presence of abnormally large observations, and even a single outlier can significantly change the value of the estimate, which manifests itself by increasing the bias. In order to reduce this effect, a function ψ with a different kind of monotony is needed. According to [45], the optimal choice of the robustifying function ψ is based on the analysis of the asymptotic variance of the estimator. It can be shown that the asymptotic variance of an M-estimator defined by (4.61) at a distribution F from the class (4.60) is given by [43]:

$$V(\psi;F)=\frac{\int\psi^2 dF}{\left(\int\psi' dF\right)^2} \tag{4.62}$$

Moreover, the asymptotic variance satisfies the information inequality:

$$V(\psi,F)\geq\frac{1}{I(F)} \tag{4.63}$$

where $I(F)=\int\left(\frac{F''}{F'}\right)^2 dF$ is the Fisher information. The inequality is always strict, unless function ψ is proportional to $-\frac{F''}{F'}$ which means that the M-estimator is, in fact, the maximum likelihood estimator. Huber shows that the asymptotic variance (4.63) has a saddle point, that is, there is such a pair (ψ_0, F_0) that:

$$\sup_{F\in\mathbf{F}_\varepsilon} V(\psi_0;F)=V(\psi_0;F_0)=\inf_{\psi} V(\psi;F_0) \tag{4.64}$$

To find the pair, we first search for the least favorable distribution F_0, that is, the distribution that minimizes the Fisher information $I(F)$ over the class $\mathbf{F}_\varepsilon$ described by (4.60). With the least favorable distribution found, we simply choose the maximum likelihood estimator, with $\psi_0=-\frac{F_0''}{F_0'}$. In other words, the estimator minimizes the maximal asymptotic variance over the class $\mathbf{F}_\varepsilon$ and represents the well-known Huber solution of the robust estimation problem that is optimal in a minimax sense.

Huber showed that the density of the least favorable distribution minimizing the Fisher information over the class $\mathbf{F}_\varepsilon$ with unity variance of the central component ($\sigma^2 = 1$) is given by [43, 44]:

$$f_0(x)=\begin{cases}\frac{1-\varepsilon}{\sqrt{2\pi}}\exp\left(-\frac{x^2}{2}\right) & \text{for } |x|\leq k\\ \frac{1-\varepsilon}{\sqrt{2\pi}}\exp\left(\frac{k^2}{2}-k\cdot|x|\right) & \text{for } |x|>k\end{cases} \tag{4.65}$$

where k and ε are related through the equation:

$$k^{-1}\varphi(k)-Q(k)=0.5\varepsilon(1-\varepsilon)^{-1} \tag{4.66}$$

where ϕ is the density of the standard normal distribution and:

$$Q(u) \doteq \frac{1}{\sqrt{2\pi}} \int_{u}^{\infty} \exp\left(-x^2/2\right) dx \tag{4.67}$$

The corresponding robustifying function $\psi_h(x) = -f_0'(x)/f_0(x)$ is given by:

$$\psi_h(x) = \begin{cases} x & \textit{for} \quad |x| \le k \\ k \cdot \operatorname{sgn}(x) & \textit{for} \quad |x| > k \end{cases} \tag{4.68}$$

The minimax robustification presented above might not be very effective when highly impulsive noise is present. In the case of high impulsivity better results are obtained using a non-monotone robustifying function, decreasing or even having zero values for large values of the input signal.

In the CDMA communication systems to improve the multiuser detection [46], a new class of robustifying functions called α-functions has been proposed:

$$\psi_\alpha(x) = xe^{-\alpha x^2}, \quad \alpha > 0 \tag{4.69}$$

It can be noted that for large values of the argument, $\psi_\alpha(x)$ decreases exponentially so such a robustification function should be very effective for impulsive noise suppression. To make use of the function, an appropriate value of the parameter α has to be chosen. The way of choosing this value, presented in [46], is based on comparing the asymptotic variance of the estimation error corresponding to α-function with that of minimax robustification. Both variances are determined by (4.64) with an appropriate choice of the robustifying function ψ. These variances will be denoted as $V_\alpha(\varepsilon, \kappa)$ and $V_H(\varepsilon, \kappa)$ for the α-function (4.67) and the Huber minimax function (4.65), respectively. In [46] the distance measured was defined as:

$$D_H(\alpha) = \int_{\varepsilon_1}^{\varepsilon_2} \int_{\kappa_1}^{\kappa_2} \left[V_\alpha(\varepsilon,\kappa) - V_H(\varepsilon,\kappa)\right]^2 d\varepsilon d\kappa \tag{4.70}$$

where integration intervals are selected so as to cover the range of variation of the contamination noise parameters that are of interest to us. It is shown in [46] that $D_H(\alpha)$ has a unique minimum. For a reasonable range of the contamination noise parameters, that is, $0.01 \le \varepsilon \le 0.1$ and $1 \le \kappa \le 100$, the minimum is achieved at $\alpha = 0.15$.

In the classical problem of the robust estimation of a location parameter, the minimax optimality is very often given up in order to obtain an estimator with better performance for outliers with very large magnitudes. In this case, a non-monotone function ψ, vanishing outside some interval around zero is preferred. Among many functions with this feature [43], the Tukey bisquare function (also called the biweight function) is very popular in literature. It has the following form:

$$\psi_{bs}(x) = \begin{cases} x\left[1-\left(\frac{x}{a}\right)^2\right]^2 & \text{for} \quad |x| \le a \\ 0 & \text{for} \quad |x| > a \end{cases} \tag{4.71}$$

In the estimation problem of a location parameter, the tuning constant a is chosen to ensure the desired level of asymptotic efficiency in the Gaussian case. For example, $a = 4.68$ guarantees 95% of maximal asymptotic efficiency achieved in the Gaussian case by a linear estimator.

After the theoretical consideration it is time to show the shape of the robustifying functions. Figure 4.40 presents the robustifying functions defined by (4.68), (4.89), and (4.71) for a selected value of parameters. In the case of functions $\psi_b(x)$ described by (4.68) and $\psi_{bs}(x)$ described by (4.71), the increase of characteristic parameters k and a causes the linear or pseudo-linear region of the functions to widen. The behavior of the $\psi_\alpha(x)$ function given by (4.89) is the opposite; an increase of α narrows the pseudo-linear region of the function. For that reason, the reciprocal of α is used in further experiments for scaling the x-axis.

The presented functions have been developed for real signals. In the case of noise radars all signals are complex signals after converting to the baseband, and the theory presented above is not directly applicable. However, it can be observed [47–51] that even a simple and nonoptimal method of robustification can improve the radar detection in the case of heavy tail distributed noise. So for noise radar robustification it is assumed that the robustification function is applied to the magnitude of the received signal $x_R(t)$, while the phase of the signal is preserved:

$$y_R(t) = \psi(|x_R(t)|)\exp(j\arg(x_R(t))) \tag{4.72}$$

Afterwards a nonlinear processing signal is passed to the correlation processing described by (3.7). The quality measure of the robustification is the S/N after the correlation procedure, calculated as the ratio of the maximum peak to the mean noise floor [47]. The properties of robustification are related to two main parameters: the shape (or shape parameter) of the robustifying function

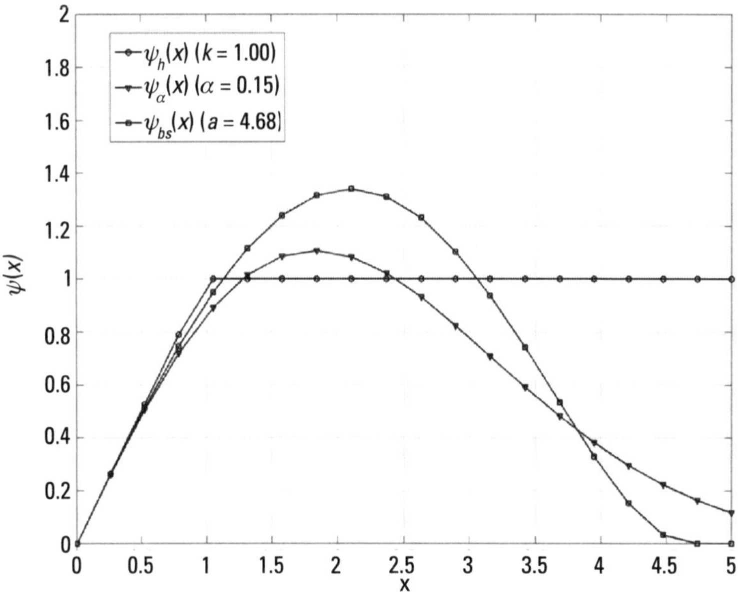

Figure 4.40 Example shapes of robustification functions.

and the parameter (or power) of input noise. In real situations noise parameters are usually unknown, but can easily be measured (e.g., input noise power). As a result, the robustifying function parameter can be modified (or rather normalized) accordingly to the measured noise power. The research on robust detection in noise radar presented by the author in [47, 48] shows that the normalization parameter should be proportional to the amplitude of the central Gaussian component (the first component in (4.60)) of the input noise. The value of the parameter can be determined in various ways. One of the simplest choices, well fitted to the Gaussian case, is the standard deviation (or the power):

$$P_0 = \sigma\left(X_R(t)\right) \tag{4.73}$$

However, the use of the standard deviation of the input signal for signal normalization may lead to unexpectedly high results, as the standard deviation will change significantly with varying parameters of the impulsive component of the noise. For this reason, additional normalizing methods will be presented, which are more robust to parameter changes. Among many possibilities, robust estimators will be shown. The first is the mean value of the absolute value of the signal:

$$P_1 = E\left[\left|X_R(t)\right|\right] \tag{4.74}$$

That value can easily be calculated, while the summation is relatively cheap from a numerical point of view. The problem is that we usually have real (I) and imaginary (Q) parts of the received signal. Further investigation shows that it is possible to sum the absolute values of the I Q components instead of summing the absolute values of the signal.

The second candidate for the robust estimator is the exponent of the mean value of the logarithm of the absolute value of the signal:

$$P_2 = \exp\left[E\left(\log\left|X_R(t)\right|\right)\right] \tag{4.75}$$

As in the previous case the computational complexity is moderate as it is possible to use a look-up table to calculate the log function, and again instead of calculating the abs of the signal, we can sum the abs of the I Q components.

The third candidate is the median value of the absolute value of the signal:

$$P_3 = med\left(\left|X_R(t)\right|\right) \tag{4.76}$$

The calculation of the median value is very intensive computationally as it requires the sorting of a long data record.

In the examples presented below the input noise was modelled using epsilon-contaminated Gaussian white noise with the probability of contamination and the effective amplitude of contamination $\sqrt{\kappa}$. The first problem is knowing how robust the estimators (4.73) through (4.76) of the uncontaminated noise component are. The results provided by the estimators mentioned above versus noise parameters ε and κ are presented in Figures 4.41 and 4.42. It can be seen that increasing noise impulsiveness by increasing ε or κ causes a significant increase of the estimator P_0 based on signal variance. The estimator based on the mean amplitude P_1 is less sensitive to contamination, but still its value changes about four times in the presented parameter range. Much better results can be obtained with an estimator based on the geometrical mean amplitude P_2 (mean log value of noise). It varies below two times and only for a very high prob ability of contamination (above 20%). The estimator based on the median of the input amplitude P_3 remains fairly constant versus ε and κ but it is a more costly estimator. However, several iterative algorithms exist which are based on a simple comparison and summation, and can estimate the median of long signal vectors with sufficient accuracy. The computational complexity of those algorithms is of the order of N rather than $N\log N$ (required by a fast sorting algorithm) and does not need to remember the whole signal vector.

To show the importance of the problem, a simple calculation of the expected signal losses can be made. In the classical Gaussian case the S/N after correlation processing is improved by the time-bandwidth product. In the con-

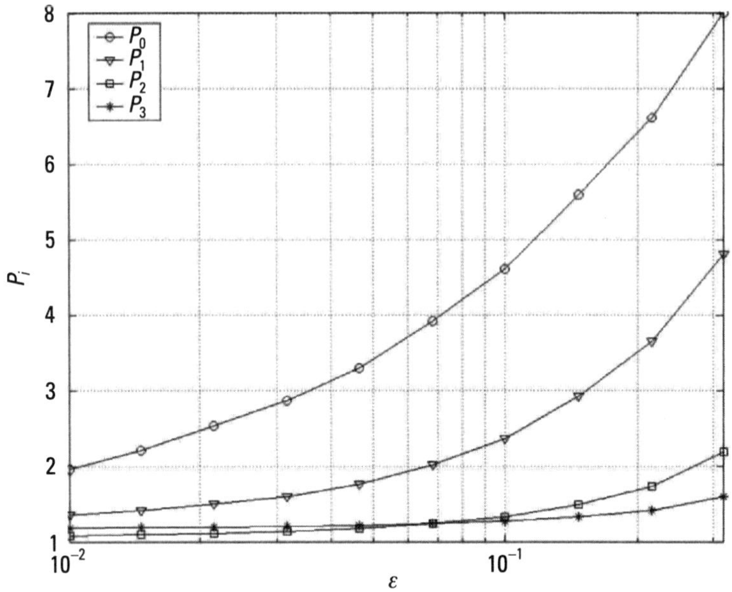

Figure 4.41 Value of parameter Pi versus ε ($\kappa = 100$).

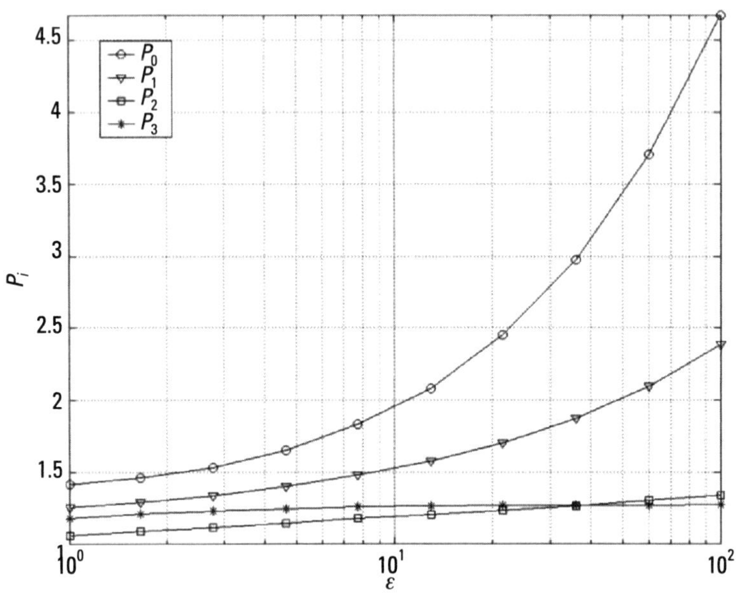

Figure 4.42 Value of parameter Pi versus κ ($\varepsilon = 0.1$).

tamination case the signal level after correlation is unchanged, but the noise level is increased from the original noise level P_N to the value:

$$P_{NC} = P_N\left((1-\varepsilon)+\varepsilon\kappa\right) \quad (4.77)$$

The example of a signal to noise losses, normalized to an S/N value without contamination is presented in Figure 4.43. As it can be seen, up to 20-dB losses can be expected for 10% of a strong contamination. In most cases expected losses will be between 2 and 10 dB.

The robustification function presented in Figure 4.40 can be divided into two major classes. To the first class belong all the functions that resulted in zeroing the outlier [as described by (4.69) and (4.71)]. If the zeroing is perfect, then each outliers will be substituted by the zero value, canceling both the outlier and the signal. The effective integration time is in this case reduced by the factor 1-ε and the integration losses (in decibels) can be estimated as:

$$I_{LZ} = -10 * \log_{10}(1-\varepsilon) \quad (4.78)$$

To the second class of the robustification function belong all functions that result in the saturating of the outlier value at the selected threshold. Saturation limits the outlier power but attenuates significantly the signal. In a simplified consideration, we can assume that in all saturated samples the signal is completely cut out, and the noise remains at the threshold value. Setting the threshold level close to the mean value of the noise, we can estimate that the losses will be two to three times higher than predicted by (4.78). The value of

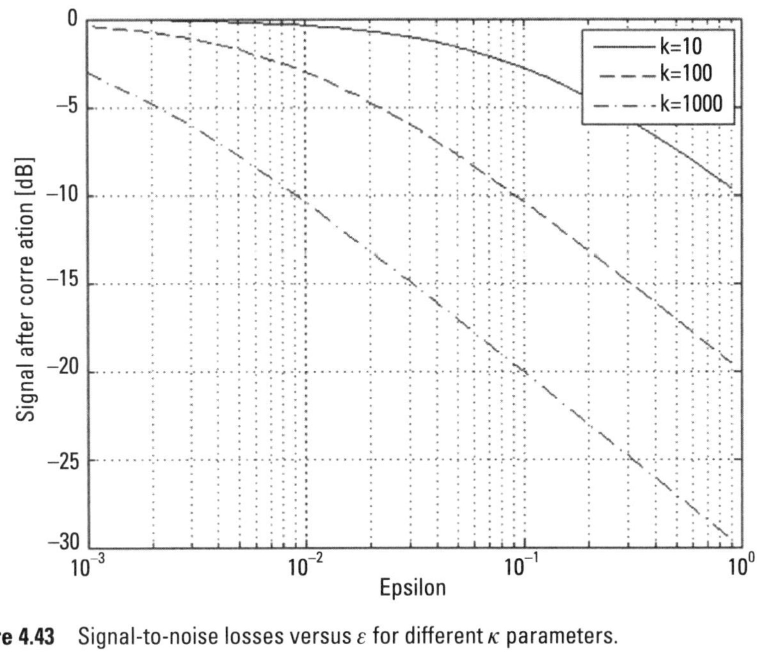

Figure 4.43 Signal-to-noise losses versus ε for different κ parameters.

the predicted losses versus the contamination intensity ε for both classes of robustifying functions is plotted in Figure 4.44.

As it is shown in Figure 4.44, the losses for 10% of the contamination are predicted at the 0.5-dB level for the first class of robustification functions (cutting the outliers) and to 1.5 dB for the second class of robustification functions (saturation of outlier). The analyses presented above are based on very high simplification, so more detailed analyses are required to verify their properties for a different level of contamination.

The proposed robust detection of weak echoes in the noise radar was verified with Monte Carlo simulations. The transmitted signal envelope was simulated as complex Gaussian noise. The received signal after conversion to the baseband was modeled as the sum of the delayed and Doppler shifted copy of the transmitted signal (target echo model) and the external noise. The input noise containing impulse distortion was modeled as the ε-contaminated white complex noise. The input Gaussian noise component was generated in the same manner as the transmitted signal, and the impulse noise component was modelled by multiplying randomly selected noise samples (with probability ε) by the factor $\sqrt{\kappa}$. The input S/N was set to −12 dB for $\varepsilon = 0$ (pure Gaussian input noise) to minimize the influence of the echo signal on the input noise probability density function. This assumption is very realistic since the strongest echoes can be easily detected and removed without robustification. After the cleaning operation, the S/N is usually much lower than 0 dB.

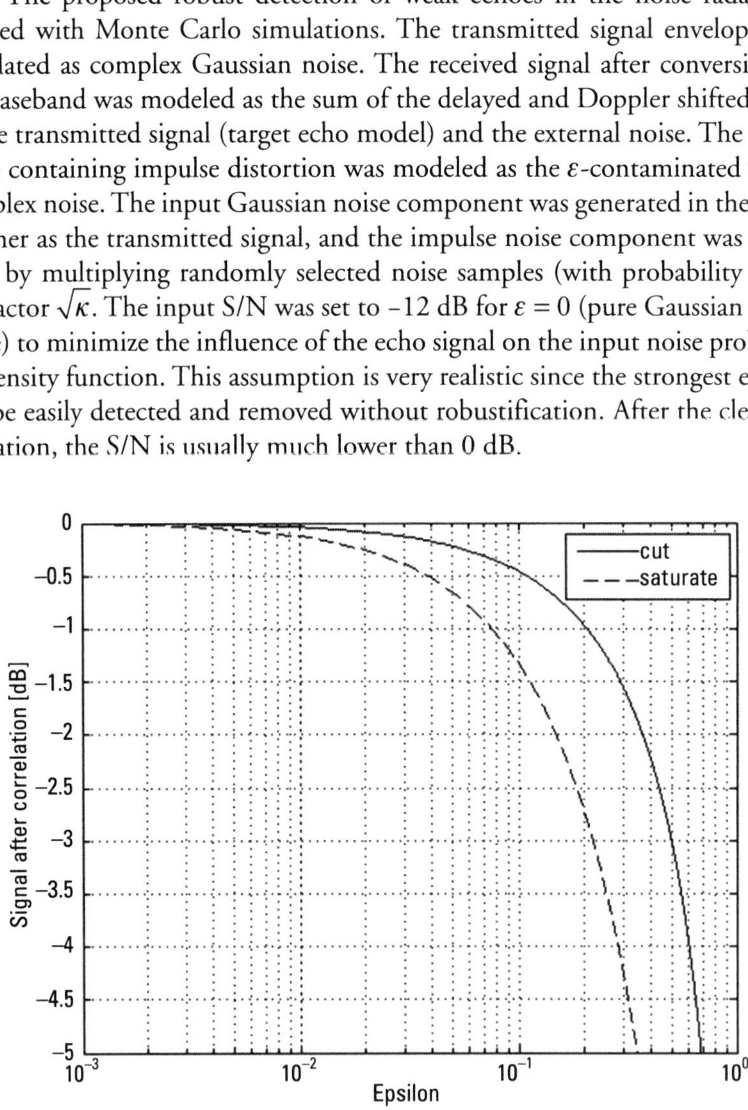

Figure 4.44 Signal-to-noise losses versus ε for two classes of robustification functions.

The effectiveness of three previously described robustification functions are shown in Figure 4.45. The simulated impulsive noise had the following parameters: $\kappa = 100$ and $\varepsilon = 0.1$. The dash-dotted line shows the S/N level for ε-contaminated noise without any robustification. The solid line shows the S/N level obtained for Gaussian noise with no robustifying function. The detection losses without robustification are at the level of 10 dB, which is in agreement with (4.78) (see Figure 4.43).

It can be seen that by choosing the characteristic parameter of a function appropriately, a performance comparable to the Gaussian case can be obtained (the worst results were obtained for the Huber function ψ_h with a 1.5-dB loss in comparison to the Gaussian case, which again is in good agreement with the prediction shown in Figure 4.44). The result was achieved for the two other robustification functions which canceled the outlier, but the losses are within 1 dB. This means that the prediction described by (4.78) is slightly too optimistic and the losses are almost twice higher than predicted. This is because in the derivation of (4.78) the assumption of the perfect cancellation of outliers was taken, while in practice two effects occur: the outliers are not fully canceled and the high values of the signal are attenuated, except for the robustification function.

In the example presented, a single set of parameters κ and ε was selected. In the practical case it would be advantageous to find such a set of parameters

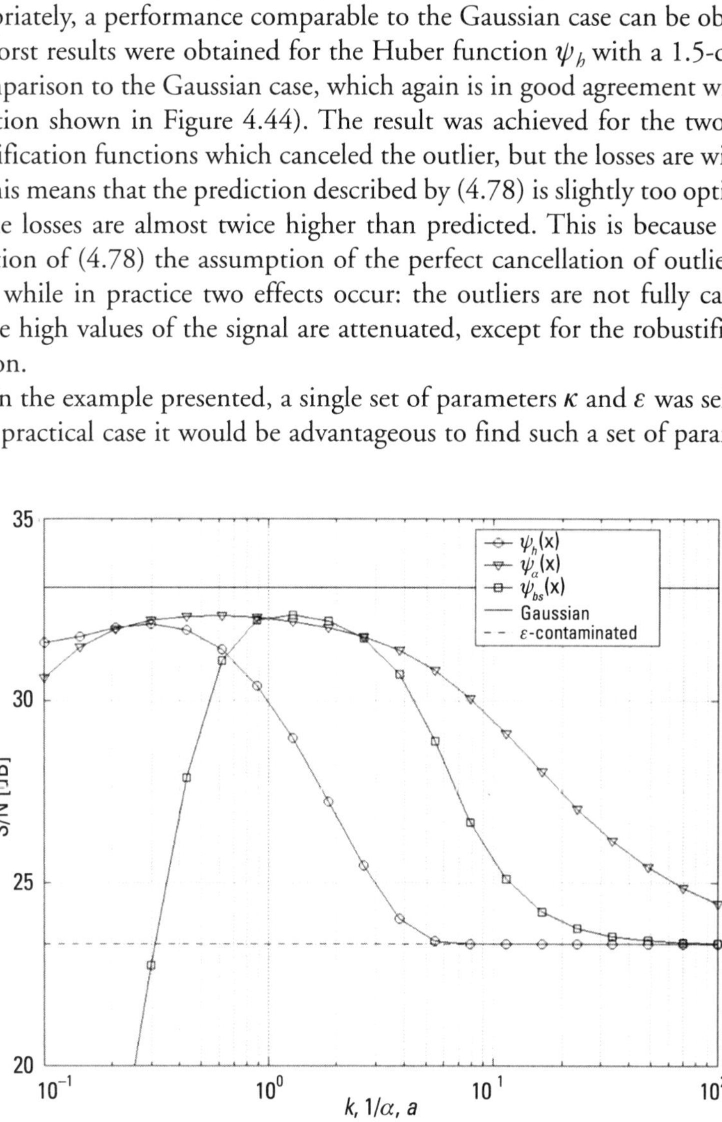

Figure 4.45 S/N after correlation versus characteristic parameters for three robustifying functions.

that could give satisfying results for all possible (expected) values of κ and ε. The alternative approach would lead to the estimation of the values of κ and ε and the choosing of the appropriate robustifyiung parameter. As the optimal S/N values after correlation are almost the same for the robustification functions $\psi_\alpha(x)$ and $\psi_{bs}(x)$, and the function $\psi_\alpha(x)$ is less sensitive to an improper choice of the robustification parameter than function $\psi_{bs}(x)$, the function $\psi_\alpha(x)$ was chosen for further presentation.

As mentioned before, the parameter of the robustifying function is strongly related to the choice of the normalising parameter P. Also stated before, the classical power estimator based on the standard deviation of the input signal may not be the best one since it will vary with changes in the noise characteristics. As a result, the optimal value of the parameter α changes significantly with the change of the input noise parameters ε and κ. This situation is illustrated in Figures 4.46 and 4.47, showing the S/N versus α for different ε and κ. For higher impulsiveness, the optimal value of $1/\alpha$ is smaller and it is not possible to choose a single value of α that will significantly improve detection for all possible values of ε and .

The situation is much better when the robust estimator of noise power based on the median value of the input noise amplitude is selected, as discussed before. Figures 4.48 and 4.49 show the S/N versus α parameter for different ε and κ in the case where the robust median-based parameter P_3 is used for normalization. The curves have approximately the same behaviour, irrespective

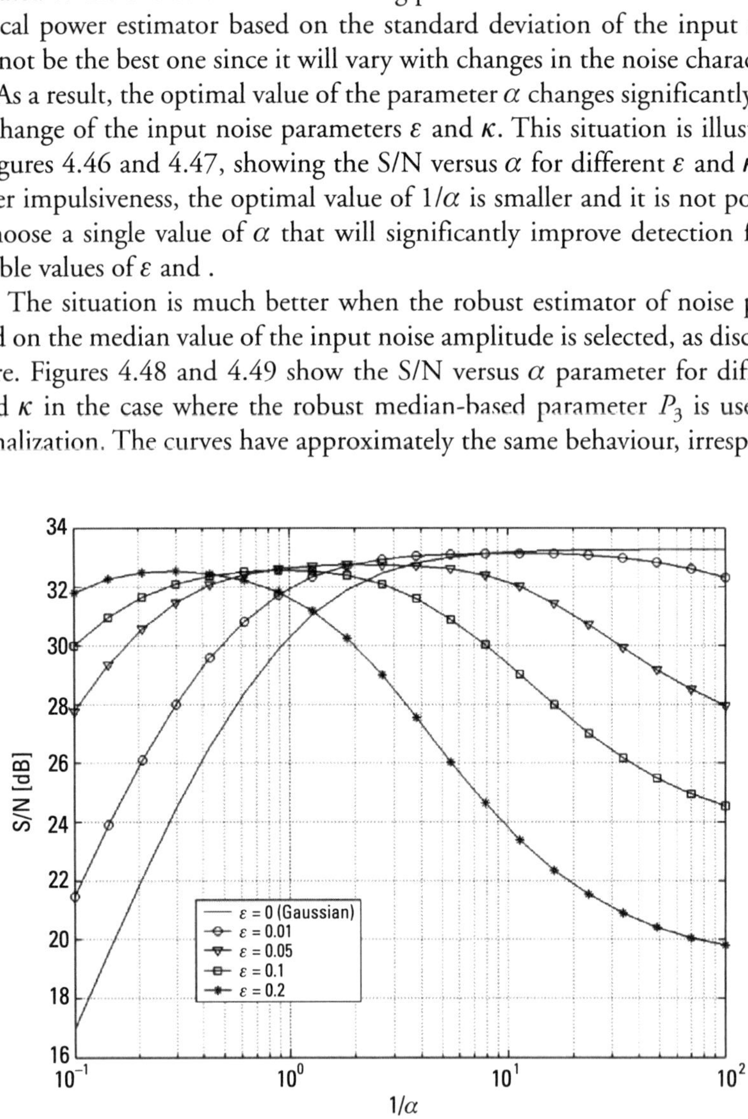

Figure 4.46 S/N versus parameter α with normalization using P_0 for different ε ($\kappa = 100$).

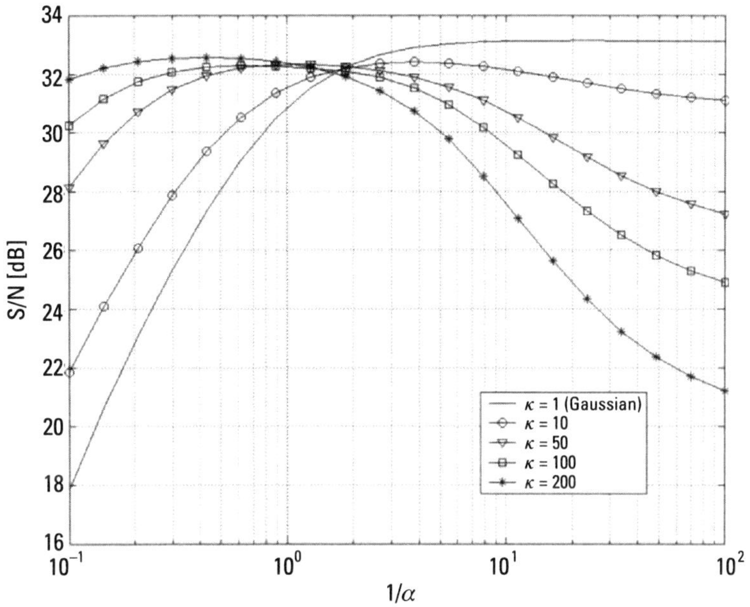

Figure 4.47 S/N versus parameter α with normalization using P_0 for different κ ($\varepsilon = 0.1$).

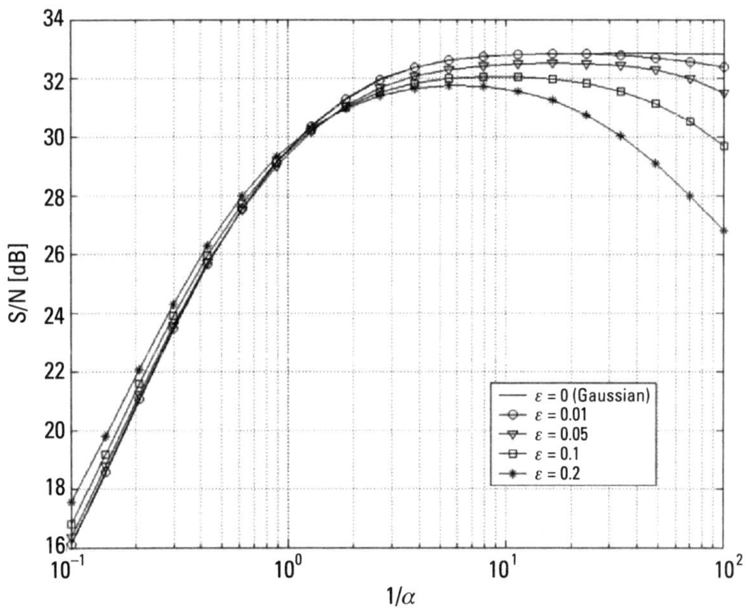

Figure 4.48 S/N versus parameter α with normalization using P_3 for different ε ($\kappa = 100$).

of the impulsiveness of the noise. This allows us to choose one value of α, in this example equal to 0.2 ($1/\alpha = 5$), which yields acceptable results for a wide

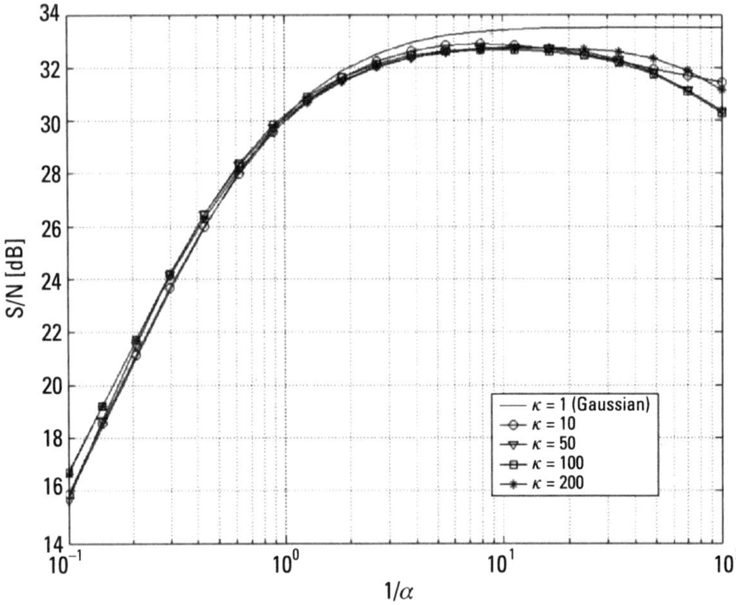

Figure 4.49 S/N versus parameter α with normalization using P_3 for different κ ($\varepsilon = 0.1$).

range of variations of the impulsiveness of the noise. Detection losses in the case of pure Gaussian noise are smaller than 0.3 dB and the difference between the optimal value of S/N and the actual S/N value for any particular values of the ε and κ parameters does not exceed 1 dB. It is possible to further reduce the losses for the Gaussian noise by decreasing the α parameter to 0.1, but the losses for a high value of ε will increase in that case.

The robustification requires very low input signal to noise ratio, much lower than 0 dB (usually in the range of −35 to −10 dB). If the input signal to noise level is much higher (above 0 dB), then the dominant component will be echo, not noise, but after classical correlation the signal-to-noise level will be higher than the time bandwidth product, usually above 50 dB. In this case even 20-dB losses will not significantly influence the detection process.

The question is how to operate in an environment where both strong (such as ground clutter and nearby targets) and weak (far) targets exist simultaneously. The proposed processing structure is presented in Figure 4.50.

The signal received by the receiving antenna, after amplification and conversion to the baseband, is passed to the adaptive ground clutter canceler, where all stationary or low-speed components are removed. The clutter-free signal is then passed to the range-Doppler correlator, where all strong targets are detected and their parameters like amplitude, range, range velocity, and acceleration are estimated. In having those estimates, it is possible to pass the signal to a strong echo canceler, where the signal is CLEANed of all strong components.

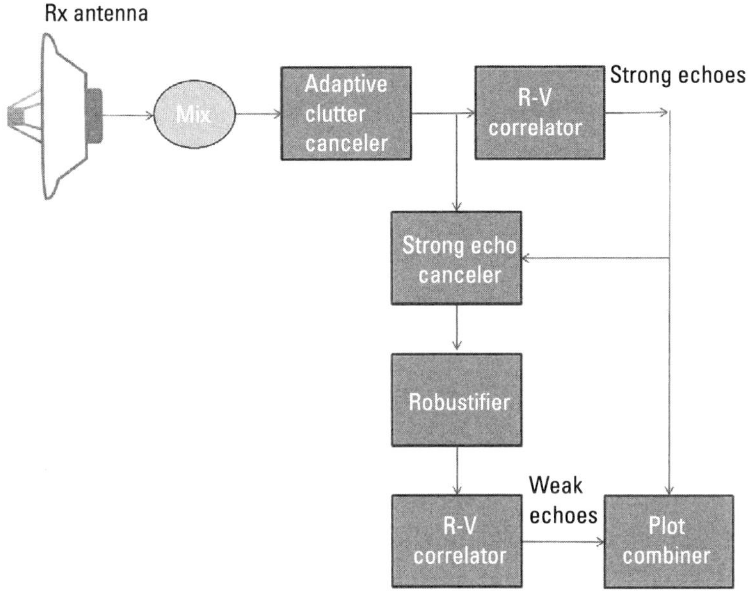

Figure 4.50 Proposed signal processing block diagram of the noise radar.

After this CLEANing stage, the signal consists only of thermal noise, interferences, and weak echoes buried in the noise. Such a signal is being robustified using a nonlinear transform and passed to the final range-Doppler correlator where the weak targets are detected and their parameters are estimated. All extracted targets (plots) are then passed to the plot combiner and further on to a data processing unit (not shown in the picture) where final data association and target tracking is performed.

To simplify the picture, the reference signal paths are not plotted, but it is obvious that in most processing parts a reference signal is used for CLEANing and for correlation.

Some more remarks on such combined processing are presented in [47–49] where as stated the signal CLEANing is used prior to robustification. The result of applying the robustification function in the case of heavy ground clutter is presented in Figure 4.51. The clutter power was approximately 25 dB above the noise level and 50 dB above the moving target echo power level. The results obtained in both cases (with and without clutter, as depicted in Figures 4.51 and 4.48, respectively), are almost identical. This is proof that the robustifying procedure can also be successful in the case of heavy clutter and strong targets.

The performance of the noise radar based on a linear correlator can drop dramatically with impulse noise and interferences. In a classical pulse radar, impulse interferences, which are a potential source of false detections, are canceled

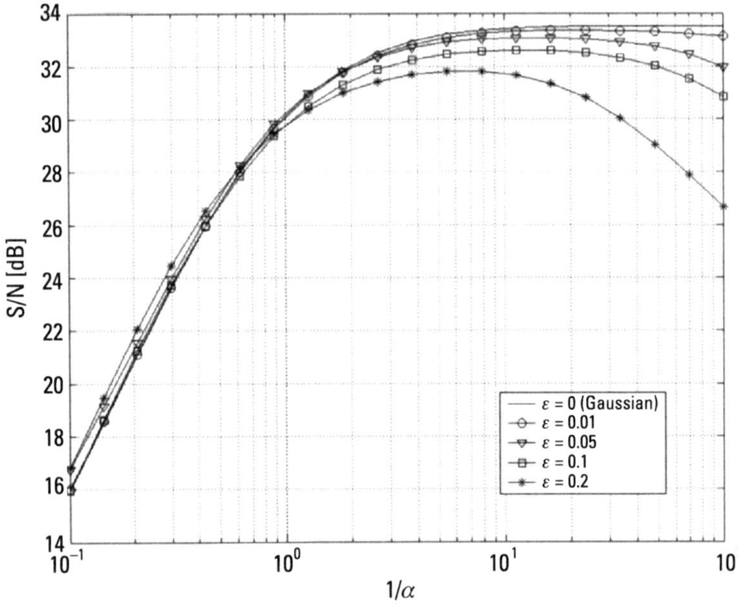

Figure 4.51 S/N versus parameter α with normalization using P_3 for different ε (κ = 100). The received signal contains strong ground clutter that was removed with a clutter removal procedure prior to robustification.

by nonlinear procedures such as binary integration. In a noise radar, impulse interferences do not produce false targets but increase the noise floor and thus decrease the radar sensitivity. The presented procedure of noise radar robustification restores the sensitivity of noise radar almost to the level where no non-Gaussian interferences are present, at the cost of no more than 0.5-dB losses for the Gaussian case and no more than 1-dB losses in the non-Gaussian case. The presented procedure shows its high performance in the presence of heavy clutter or strong echoes, which have to be removed prior to nonlinear signal transformation.

References

[1] Kulpa, K. S., Z. Czekala, and M. Smolarczyk, "Long-Integration-Time Noise Surveillance Radar," *First International Workshop on the Noise Radar Technology (NRTW-2002)*, September 18–20, 2002, pp. 238–243.

[2] Kulpa, K. S., and Z. Czekala, "Masking Effect and Its Removal in PCL Radar," *IEE Proceedings Radar, Sonar and Navigation*, Vol. 152, No. 3, June 3, 2005, pp. 174–178.

[3] Kulpa, K., and Z. Czekala, "Short Distance Clutter Masking Effects in Noise Radars," *J. Applied Radio Electronics*, Vol. 4, No. 1, Kharkov, Ukraine, 2005, pp. 96–98.

[4] Howland, P. E., D. Maksimiuk, and G. Reitsma, "FM Radio Based Bistatic Radar," *IEE Proceedings Radar, Sonar and Navigation,* Vol. 152, No. 3, June 3, 2005, pp. 107–115.

[5] Kulpa, K. S., and Z. Czekała, "Ground Clutter Suppression in Noise Radar," *Proc. Int. Conf. RADAR 2004,* Toulouse, France, October 18–22, 2004, p. 236.

[6] Kulpa, K., and Z. Czekała, "Long-Range Performance Increase in Passive PCL Radar," *3rd Multinational Conference on Passive and Covert Radar (PCR-2003),* University of Washington Applied Physics Laboratory, Seattle, WA, October 21–23, 2003.

[7] Segalovitz, A., and B. D. Frieden, "A 'CLEAN'-Type Deconvolution Algorithm," *Astron. Astrophys.*, Vol. 70, 1978, pp. 335–343.

[8] Tsao, J., and B. D. Steinberg, "Reduction of Sidelobe and Speckle Artifacts in Microwave Imaging: The CLEAN Technique," *IEEE Transactions on Antennas and Propagation,* Vol. 36, April 1988, pp. 543–556.

[9] Nałcz, M., and K. Kulpa, "Range and Azimuth Estimation Using Raw Data in DSP-Based Radar System," *Proc. 12th Int. Conf. on Microwaves and Radar MIKON-98,* Krakow, Poland, Vol. 3, May 20–22, 1998, pp. 871–875.

[10] Haykin, S., *Adaptive Filter Theory,* 3rd ed., Englewood Cliffs, NJ: Prentice-Hall, 1991.

[11] Kulpa, K., "Adaptive Clutter Rejection in Bi-Static CW Radar," *Proc. Int. Radar Symp. IRS-2004,* Warsaw, Poland, May 2004, pp. 61–66.

[12] Misiurewicz, J., and K. Kulpa, "Stretch Processing for Masking Effect Removal in Noise Radar," *IET Radar, Sonar & Navigation,* Vol. 2, No. 4, August 2008, pp. 274–283.

[13] Thevenaz, P., T. Blu, and M. Unser, "Interpolation Revisited," *IEEE Transactions on Medical Imaging,* Vol. 19, July 2000, pp. 739–758.

[14] Oppenheim, A. V., R. W. Schafer, and J. R. Buck, *Discrete-Time Signal Processing,* 2nd ed., Upper Saddle River, NJ: Prentice-Hall, 1999.

[15] Evangelista, G., "Design of Digital Systems for Arbitrary Sampling Rate Conversion," *Signal Process,* Vol. 83, No. 2, 2003, pp. 377–387.

[16] Schnitter, P., "Polyphase Resampling with Arbitrary (Non-Rational or Time-Varying) Rate," http://cnx.org/content/m10456/latest, October 2005, accessed August 2007.

[17] Frigo, M., and S. G. Johnson, "FFTW: An Adaptive Software Architecture for the FFT," *Proc. IEEE Intl. Conf. on Acoustics, Speech, and Signal Processing,* Vol. 3, Seattle, WA, May 1998, pp. 1381–1384.

[18] Misiurewicz, J., "Resampling Methods for Stretch Processing in PCL Radars," *Proc. SPIE 6937, Photonics Applications in Astronomy, Communications, Industry, and High-Energy Physics Experiments,* 2007, pp. 693–734.

[19] Li, A., K. Mueller, and T. Ernst, "Methods for Efficient, High Quality Volume Resampling in the Frequency Domain," *IEEE Visualization 2004,* Austin, TX, October 10–15, 2004, pp. 3–8.

[20] Strang, G., *Introduction to Linear Algebra,* 4th ed., Wellesley, MA: Wellesley Cambridge Press, 2009.

[21] Malanowski, M., "Comparison of Adaptive Methods for Clutter Removal in PCL Radar," *Proc. of International Radar Symposium 2006*, Krakow, Poland, May 24–26, 2006, pp. 237–240.

[22] Takaba, K., "Projection Method for Solving a Singular System of Linear Equations and Its Applications," *Numeriche Mathematik*, Vol. 17, 1971, pp. 203–214.

[23] Hayes, M. H., *Statistical Digital Signal Processing and Modeling*, New York: John Wiley & Sons, 1996.

[24] Friedlander, B., "Lattice Filters For Adaptive Processing," *Proceedings of the IEEE*, Vol. 70, No. 8, August 1982.

[25] Meller, M., and S. Tujaka, "Block Least Mean Squares Processing of Noise Radar Waveforms," *2009 IEEE Radar Conference*, May 4–8, 2009, pp. 1–6.

[26] Meller, M., and S. Tujaka, "Processing of Noise Radar Waveforms Using Block Least Mean Squares Algorithm," *IEEE Transactions on Aerospace and Electronic Systems*, Vol. 48, No. 1, January 2012, pp. 749–761.

[27] Meller, M., "Cheap Cancellation of Strong Echoes for Digital Passive and Noise Radars," *IEEE Transactions on Signal Processing*, Vol. 60, No. 5, May 2012, pp. 2654–2659.

[28] Skolnik, M. I., *Radar Handbook*, 3rd ed., New York: McGraw-Hill, 2008.

[29] Billingsley, J. B., *Low-Angle Radar Land Clutter*, Norwich NY: William Andrew, 2002.

[30] David K. Barton, *Radars: Radar Clutter*, Dedham, MA: Artech House, 1975.

[31] Axelsson, S. R. J., "Improved Clutter Suppression in Random Noise Radar," *Proc. URSI 2005 Commission F Symposium on Microwave Remote Sensing of the Earth, Oceans, Ice, and Atmosphere*, Barza d'Ispra, Italy, April 20–21, 2005.

[32] Kulpa, K., "Simple Sea Clutter Cancellation for Noise Radar," *Proc. of International Radar Symposium 2006*, Krakow, Poland, May 24–26, 2006, pp. 299–302.

[33] Daniels, D. J., "Resolution of Ultra Wide-Band Radar Signals," *Proc. IEE Radar, Sonar Navigation*, Vol. 146, No. 4, August 1999, pp. 189–194.

[34] Garmatyuk, D., and R. M. Narayanan, "Ultra-Wideband Continuous-Wave Random Noise ARC-SAR," *IEEE Transactions on Geoscience and Remote Sensing*, Vol. 40, No. 11, December 2002, pp. 2543–2552.

[35] Davis, M. E., "Technical Challenges in Ultra Wide-Band Radar Development for Target Detection and Terrain Mapping," *Proc. IEEE 1999 Radar Conf.*, Boston, MA, April 1999, pp. 1–6.

[37] Middleton, D., "Statistical-Physical Models of Electromagnetic Interference," *IEEE Transactions on Electromagnetic Compatibility*, Vol. EMC-19, No. 3, August 1977, pp. 106–127.

[38] Middleton, D., "Canonical Non-Gaussian Noise Models: Their Implications for Measurements and for Prediction of Receiver Performance," *IEEE Transactions on Electromagnetic Compatibility*, Vol. EMC-21, Vol. 3, August 1979, pp. 209–220.

[39] Middleton, D., "Procedures for Determining the Parameters of the First-Order Canonical Models of Class A and Class B Electromagnetic Interference," *IEEE Transactions on Electromagnetic Compatibility*, Vol. EMC-21, Vol. 3, August 1979, pp. 190–208.

[40] Nikias, C. L., and M. Shao, *Signal Processing with Alpha-Stable Distributions and Applications*, New York: John Wiley & Sons, 1995.

[41] Shao, M., and C. L. Nikias, "Signal Processing with Fractional Lower Order Moments: Stable Process and Their Application," *Proc. IEEE*, Vol. 81, No. 7, July 1993, pp. 986–1010.

[42] Middleton, D., "Non-Gaussian Noise Models in Signal Processing for Telecommunications: New Methods and Results for Class A and Class B Noise Models," *IEEE Transactions on Information Theory*, Vol. IT-45, No. 4, May 1999, pp. 1129–1149.

[43] Blum, R., Y. Zhang, and R. Kozick, "On the Approximation of Correlated Non-Gaussian Noise Pdfs Using Gaussian Mixture Models," *1st Conference on the Applications of Heavy Tailed Distributions in Economics, Engineering and Statistics*, American University, Washington, D.C., June 1999.

[44] Huber, P. J., *Robust Statistics*, New York: John Wiley & Sons, 1981.

[45] Huber, P. J., "Robust Estimation of Location Parameter," *The Annals of Mathematical Statistics*, Vol. 35, 1964, pp. 73–101.

[46] Wang, X., and H. V. Poor, "Robust Multiuser Detection in Non-Gaussian Channels," *IEEE Transactions on Signal Processing*, Vol. 47, No. 2, February 1999, pp. 289–305.

[47] Seyfe, B., and S. Valaee, "A New Choice of Penalty for Robust Multiuser Detection Based on M-Estimation," *IEEE Transactions on Communications*, Vol. 53, No. 2, February 2005, pp. 224–227.

[48] Kulpa, K., et al., "A Simple Robust Detection of Weak Target in Noise Radars," *European Radar Conference (EuRad) 2007*, Munich, Germany, October 8–12 2007.

[49] Kulpa, K., Z. Gajo, and M. Malanowski, "Robustification of Noise Radar Detection," *IET Radar, Sonar & Navigation*, Vol. 2, No. 4, August 2008, pp. 284–293.

[50] Malanowski, M., and K. Kulpa, "Robust Detection in Continuous-Wave Noise Radar—Experimental Results," *2010 11th International Radar Symposium (IRS)*, Warsaw, Poland, June 16–18, 2010, pp. 1–4.

[51] Lan, D., et al., "Noise Robust Radar HRRP Target Recognition Based on Multitask Factor Analysis with Small Training Data Size," *IEEE Transactions on Signal Processing*, Vol. 60, No. 7, July 2012, pp. 3546–3559.

[52] Himonas, S. D., and M. Barkat, "A Robust Radar CFAR Detector for Multiple Target Situations," *Proceedings of the 1989 IEEE National Radar Conference*, March 29–30, 1989, pp. 85–90.

5

Multistatic Noise Radar

The concept of noise radar can be used in both the monostatic and the bistatic configuration. In the monostatic configuration, as stated in Chapters 3 and 4, the transmitting and receiving antennas are placed at the same point. Theoretically, it is possible to use the same antenna for both transmission and reception, but due to very high crosstalk between the transmitting and receiving channels, such a configuration in a continuous-wave radar is used very rarely. Much more popular is the configuration with two vertically or horizontally stacked antennas. Although there is some spatial separation between the antennas, such a configuration is defined as monostatic[1]. To obtain much smaller crosstalk between the transmitting and the receiving antenna, it is possible to place those antennas in different locations, which leads to the bistatic configuration. Spatial separation of antennas has several advantages. In addition to good electromagnetic separation, it is possible to have much larger surveillance volume than in the monostatic case and to exploit high bistatic reflectivity of many man-made objects [1] as well as mitigate the problem of detection of stealth targets [2]. The radar cross-section of the target rises significantly when the transmit antenna, receive antenna and the target are almost collinear [3–7]. In this case the target is in the forward scattering region and the radar cross-section of the target can be 20–40 dB higher than it is in the monostatic case. Besides many advantages, the bistatic configuration also has some drawbacks. In the area where the forward scattering effect exists, both the range and the Doppler resolution are significantly degraded. To perform correlation processing, it is necessary to acquire the reference signal. This can be performed by using a separate reference antenna placed at the receiving site, by using a reference antenna placed

1. According to the IEEE definition, bistatic radar is a radar using antennas for transmission and reception at sufficiently different locations so that the angles or ranges from those locations to the target are significantly different.

near the transmitter, or by sending the reference signal (usually in the digital form) using a dedicated transmission channel. In the first case, when a reference signal is received by the reference antenna placed at the receiver site, there usually exists a blind zone in the direction of the transmitter where target detection is almost impossible in spite of the large increase in target cross section. To overcome this drawback, it is possible to extend the radar configuration from bistatic to multistatic, adding additional transmitting and receiving sites. In the multistatic configuration it is possible to obtain good range and Doppler resolution in the whole surveillance space and avoid or minimize blind zones, also increasing the detection probability [8].

5.1 Bistatic Configuration

The classical bistatic configuration is presented in Figure 5.1. The sector or omnidirectional transmitter illuminates the surveillance space and the receiver site. The bistatic receiver consists of two channels. The first, the surveillance (or measurement) channel, receives the echoes from moving targets. The second, the reference channel, receives the direct signal required for correlation processing and clutter rejection.

The spatial separation of the transmitter and receiver has several advantages. The first significant advantage is the possibility of better attenuation of the direct signal, so the required receiver dynamic range can be much smaller. Assuming that there is a direct line of sight between the transmitting and receiving (measurement t) antennas, then the power P_D of the direct signal received by the measurement antenna can be described by the formula:

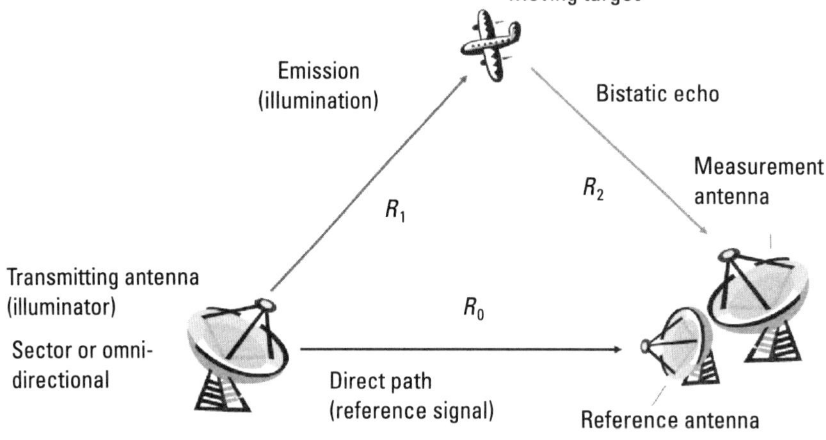

Figure 5.1 Bistatic configuration of a noise radar with direct (reference) signal channel.

$$P_D = \frac{P_T G_{TD} G_{RD} \lambda^2}{(4\pi)^2 R_0^2} \tag{5.1}$$

where P_T is the transmitted power, G_{TD} is the transmitted antenna gain in the direction of the receiver, G_{RD} is the receiver antenna gain in the direction of the transmitter, R_0 is the distance between the transmitter and receiver, and λ is the wavelength. For example, for a bistatic X-band radar with a 3-dB transmitting gain, a 20-dB receiving gain towards the transmitter, and 10 km of distance between the transmitter and the receiver, the direct signal attenuation is at the level of 110 dB. This is much more than what can be achieved in the monostatic configuration. The minimal required receiver dynamic range depends on the direct received power (5.1) and thermal receiver noise (2.5), and for the classical bistatic configuration it can be predicted by the equation:

$$P_D / P_N = \frac{P_T G_{TD} G_{RD} \lambda^2}{k T_R B (4\pi)^2 R_0^2} \tag{5.2}$$

Equation (5.2) does not take into consideration the signals reflected from targets close to the receiver, so the required receiver dynamic range can be slightly higher than predicted by (5.2). For the configuration mentioned above with the bandwidth $B = 1$ MHz and transmitted power $P_T = 10$ kW, the minimal required dynamic range is almost 80 dB. The received direct power versus the distance between the transmitter and receiver for the example mentioned above (10-kW transmitter, 1-MHz bandwidth, 3-dB transmitting antenna gain, and 20-dB receiving antenna gain) is depicted in Figure 5.2.

A further decrease of direct power may be accomplished by increasing the distance between the antennas, forming the receiver antenna radiation pattern with null at the direction of the transmitter or shielding/shadowing the receiving antenna. One of the possible configurations of a bistatic noise radar with the measurement antenna placed in a shadow region (behind the mountains as described in [9]) is presented in Figure 5.4. In such a case the reference antenna has to be placed in another location, with direct illumination of the transmitter, or it is necessary to acquire the direct signal in a different way (e.g., by using a separate data channel between the transmitter and the receiver).

The bistatic detection range can be predicted in a very similar manner as described in Section 3.2. In the monostatic case the received power can be predicted by formula (3.9). In the bistatic case the distance between the target and transmitter R_1 is usually different from the distance between the target and receiver R_2, and the power received by the measurement antenna can be predicted by the formula [1]:

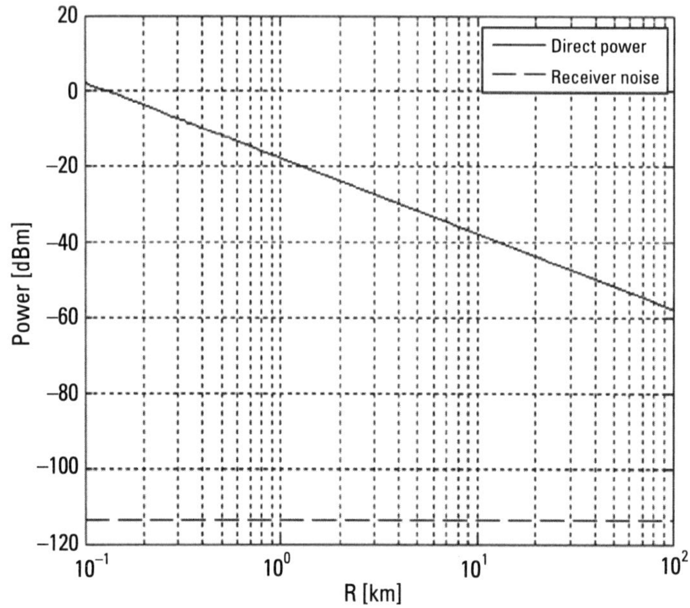

Figure 5.2 The direct received power and the noise floor of the X-band radar receiver (1-MHz bandwidth) versus the transmitter receiver baseline length.

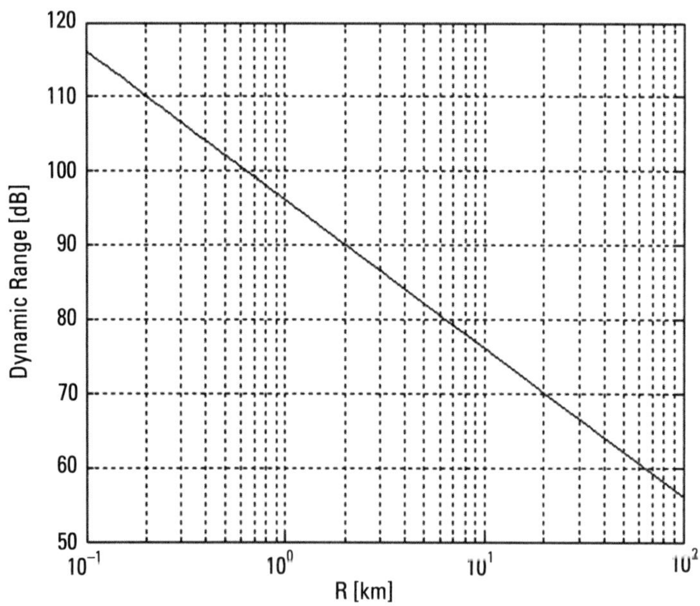

Figure 5.3 The dynamic range (direct signal to thermal noise ratio) required by the noise radar receiver versus the transmitter receiver baseline length.

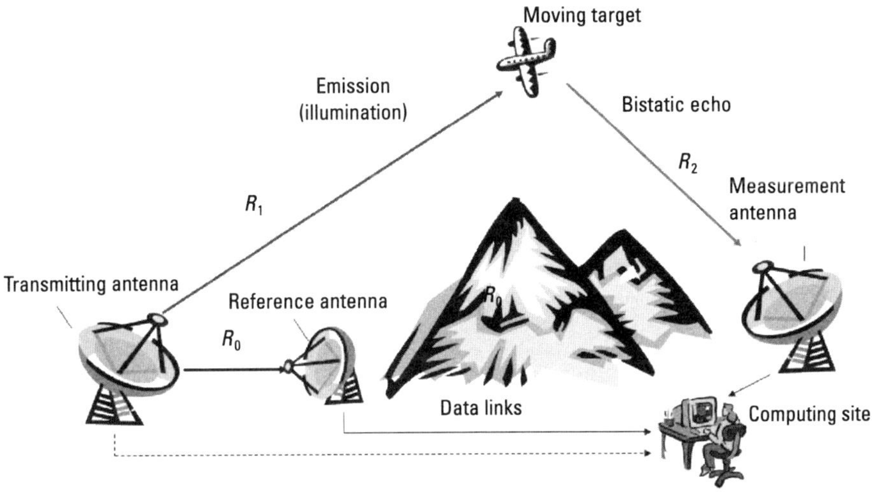

Figure 5.4 Bistatic configuration of a noise radar without direct illumination of the receiving (measurement) antenna.

$$P_R = \frac{P_T G_T G_R S_o \lambda^2}{(4\pi)^3 L R_1^2 R_2^2} \tag{5.3}$$

Comparing this power with receiver thermal noise, taking into account the gain of correlation processing and the required false alarm rate, one can obtain the detection constraint in the form:

$$\frac{P_T G_T G_R S_o \lambda^2}{(4\pi)^3 L R_1^2 R_2^2} > k T_R B \frac{D_0}{t_i B} \tag{5.4}$$

The maximum bistatic range can then be predicted using the following formula [1, 10]:

$$R_{1\max} R_{2\max} = \sqrt{\frac{P_T G_T G_R S_o \lambda^2 t_i}{(4\pi)^3 L D_O k T_R}} \tag{5.5}$$

In this case the bistatic range has two components: target-transmitter distance and target-receiver distance. For a monostatic case the coverage diagram is in the form of a circle with the radius $R_{\max}$. For a bistatic case the coverage diagram forms the Cassini oval, presented in Figure 5.5.

The theoretical coverage can be achieved only in the second bistatic configuration presented in Figure 5.4, when the direct signal cancellation is not required. However, in the first configuration, shown in Figure 5.1, it is necessary

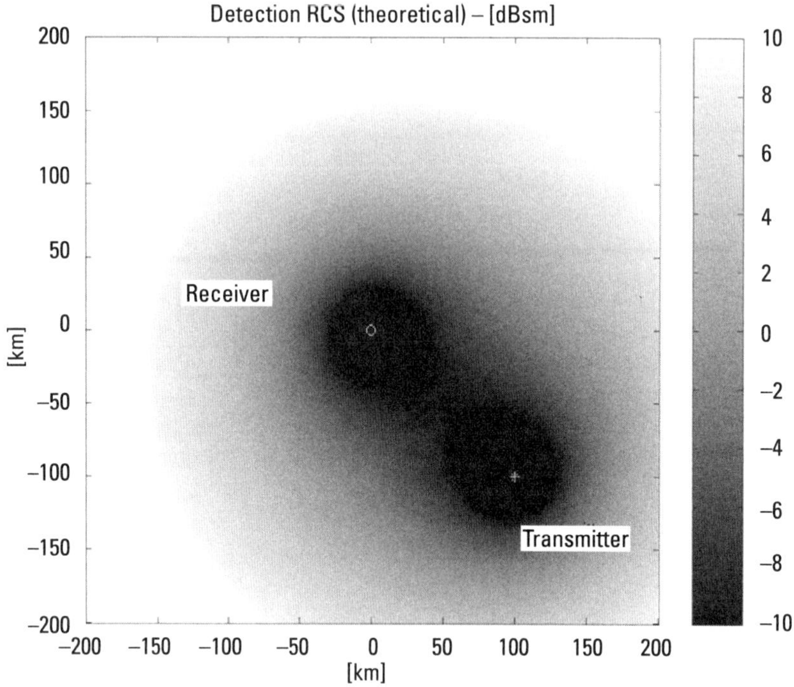

Figure 5.5 Theoretical coverage diagram of a bistatic noise radar. Grayscale: detected RCS [dBsm].

to cancel that signal. The reference antenna, apart from the direct signal, also receives the target echo. If the target is placed on the transmitter-receiver line, or close to it, then the direct signal cancellation procedure (or clutter cancellation procedure) will also cancel this target echo. The echoes of all targets present in the direct signal antenna beamwidth will be attenuated, and the attenuation factor will depend on the $G_{ref}/(G_{ref} - G_{refT})$ ratio, where G_{ref} is the reference antenna gain towards the transmitter and G_{refT} is the reference antenna gain towards the target. An example bistatic noise radar coverage diagram after the direct signal cancellation procedure is presented in Figure 5.6 [11, 12]. An antenna pattern with a main beamwidth of about 40° has been assumed in the experiment.

As one can see from Figure 5.6, it is not possible to detect the target in the direction of the transmitter and in the zone of a few degrees around this direction. The result of direct signal cancellation is the creation of a radar blind zone. The method for blind zone reduction is discussed in Section 5.1.

As in the case of the monostatic radar, the presence of strong echoes originating from the nearby target or ground clutter will reduce the detection range for simple correlation processing. The nearby point target of effective radar cross-section S_1 will produce an echo with the power:

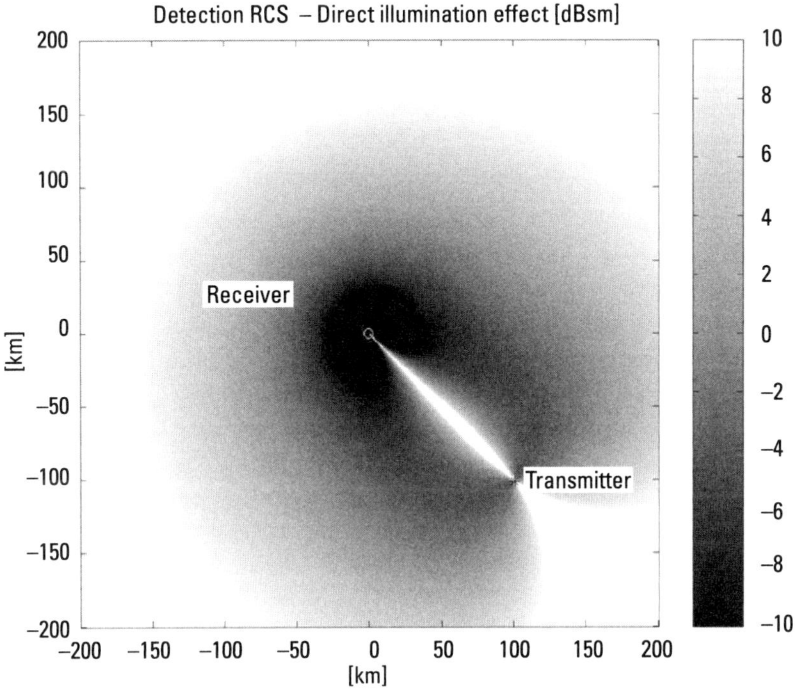

Figure 5.6 The coverage diagram of a bistatic noise radar after the direct signal cancellation procedure. Grayscale: detected RCS [dBsm].

$$P_1 = \frac{P_T G_T G_R S_1 \lambda^2}{(4\pi)^3 L R_1^2 R_2^2} \tag{5.6}$$

If this power is higher than the receiver noise power kT_RB, then the echo power will mask the weaker target echoes. The detection range losses can be calculated using the formula (5.5) substituting the original radar losses L with effective losses L_e:

$$L_e = L\left(1 + \frac{P_T G_T G_R S_1 \lambda^2}{(4\pi)^3 L R_1^2 R_2^2 k T_R B}\right) \tag{5.7}$$

The limitation of the coverage diagram due to the presence of a near, strong target is presented in Figure 5.7. In the calculation a typical receiving antenna pattern with azimuth sidelobes at the level of −25 dB has been taken into consideration. The presence of the antenna sidelobes has led to a situation in which the detection range was reduced not only in the direction of the strong

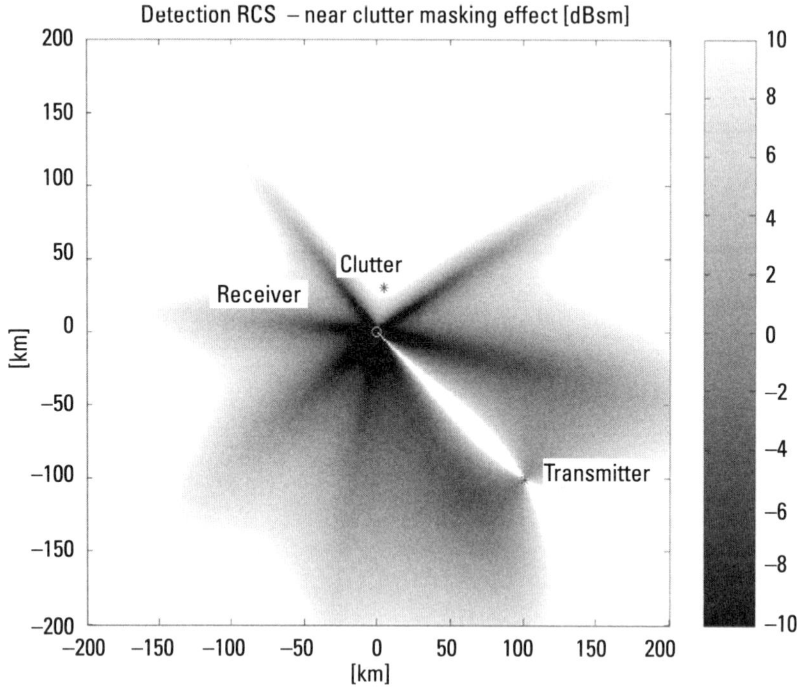

Figure 5.7 The coverage diagram of a bistatic noise radar after the direct signal cancellation procedure, when a strong nearby target (or clutter) is present. Grayscale: detected RCS (dBsm).

target, but also in the direction of the maxima of the surveillance antenna sidelobes. In the calculations presented, the assumption of the use of a multibeam antenna was made.

The masking problem can be eliminated by the use of the strong target removal procedure described in Chapter 4. The application of the procedure mentioned above can almost restore the original radar coverage [12]. In Figure 5.8 the noise radar coverage in the presence of a strong nearby echo is presented, after the clutter and strong echoes cancellation procedure.

In the case presented the strong echo was not canceled completely, but the range reduction in the direction of the strong echo is very small (less than 20%), and in the other directions the echo-free coverage has been almost fully restored.

Correlation processing in the bistatic noise radar is performed in the same manner as in the monostatic case, although the range and range velocity of the correlation peak in (3.8) now have completely different meanings. In the monostatic case the range is defined as the distance from the radar to the target. In the bistatic case the range is defined as:

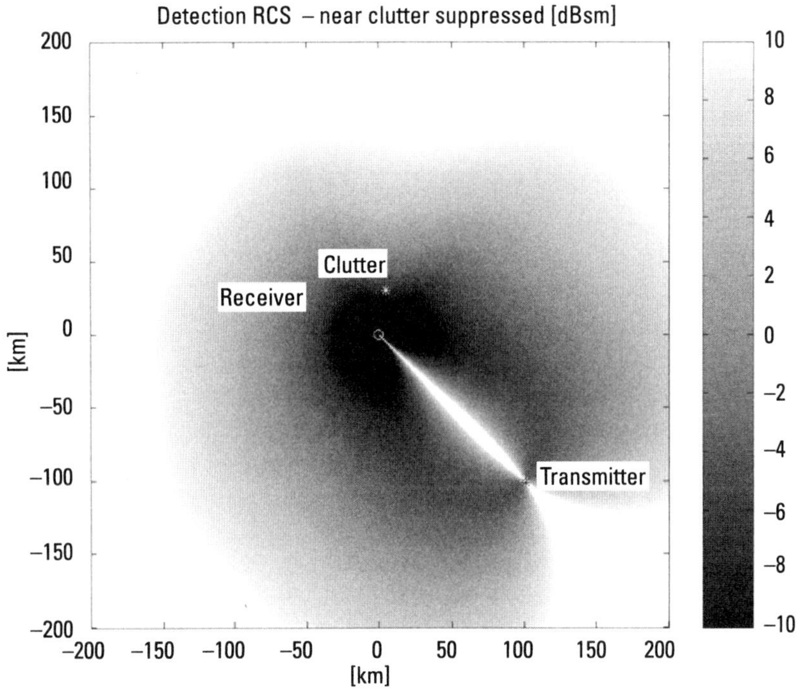

Figure 5.8 The coverage diagram of a bistatic noise radar after the clutter and strong echoes cancellation procedure, when a strong nearby target (or clutter) is present. Grayscale: detected RCS in (dBsm).

$$R = \frac{R_1 + R_2 - R_0}{2} \tag{5.8}$$

Constant range forms a circle in a two-dimensional monostatic radar and an ellipse in a bistatic radar as depicted in Figure 5.9.

Similar to the monostatic case, the angular resolution is much smaller than the bistatic range resolution and the error gate forms a long ellipse. The bistatic range is very small in the zone between the transmitter and receiver (as shown in Figures 5.10 and 5.11), so in that area the positioning error can be large.

The bistatic velocity is defined as the time derivative of the bistatic range:

$$v = \frac{dR}{2dt} = \frac{dR_1 + dR_2}{2dt} \tag{5.9}$$

If the target velocity is equal to V_T, as presented in Figure 5.10, then the bistatic velocity v can be expressed by formula [10, 13]:

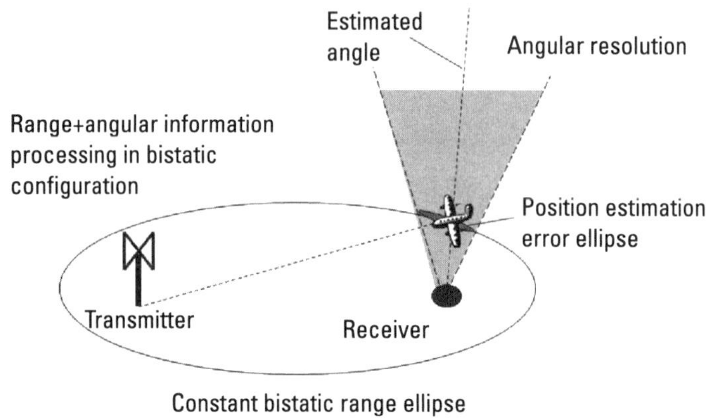

Figure 5.9 Bistatic detection of the target and target positioning error.

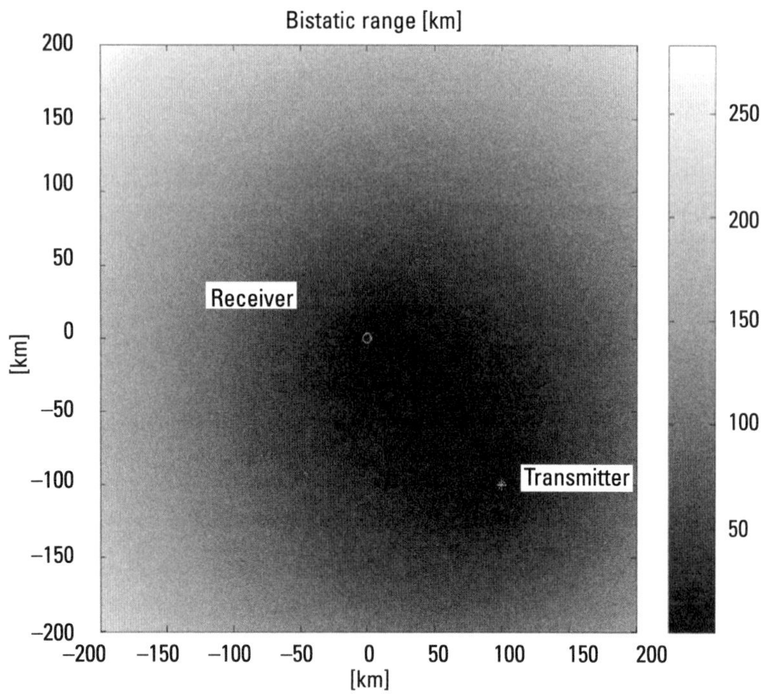

Figure 5.10 Bistatic range, grayscale in kilometers.

$$v = V_T \cos\left(\frac{\beta}{2}\right)\cos(\delta) \tag{5.10}$$

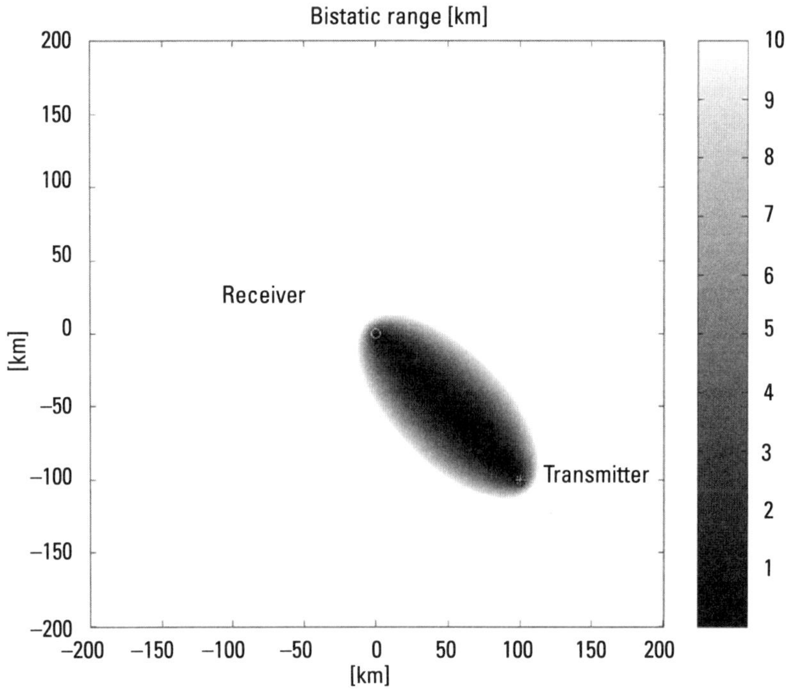

Figure 5.11 Bistatic range, grayscale in kilometers, magnification of Figure 5.10.

where β is the bistatic angle and δ is the angle between the target velocity vector and bisector of the bistatic angle β as shown in Figure 5.12.

The maximum bistatic velocity depends on the bistatic angle β and is equal to:

$$v_{\max} = V_T \cos\left(\frac{\beta}{2}\right) \tag{5.11}$$

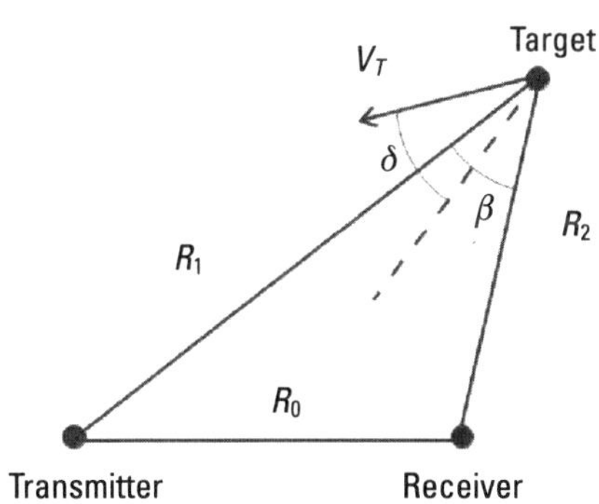

Figure 5.12 Velocity of a moving target in the bistatic configuration.

On the line between the transmitter and the receiver the bistatic angle is equal to 180°, so the maximum bistatic velocity is zero, and all targets on that line will be treated as zero-Doppler clutter. Also, targets near this line will have very small bistatic velocity and could be treated as low-Doppler clutter. The maximum bistatic velocity for a 100 m/s target is plotted in Figure 5.13.

5.2 Multistatic Configuration

The bistatic configuration, described in the previous section, has several disadvantages. It is impossible to detect a target in the blind zone between the transmitter and the receiver due to the presence of several effects. In the blind zone the target echo is attenuated significantly by the direct signal and clutter removal procedure. The maximum bistatic velocity and bistatic range in the blind zone are very close to zero, so even if it is not canceled in the process of direct signal cancellation, the target will be treated as low-Doppler near clutter.

To overcome this problem, to make the detection less sensitive to interference, to increase the positioning accuracy, and to increase the target update rate, the multistatic configuration should be used. The multistatic configuration

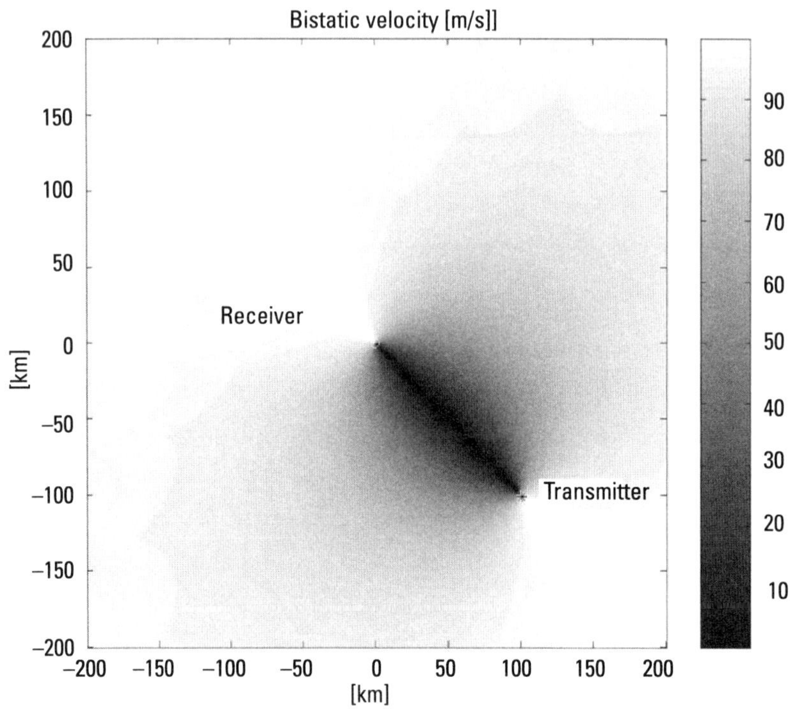

Figure 5.13 Maximum bistatic velocity for a 100 m/s target. Grayscale in m/s.

presented in Figure 5.14 may consist of one transmitter and a multiple receiver, one receiver and multiple transmitters, and a net of multiple receivers and multiple transmitters [8, 13].

In the case of multiple transmitters, it is possible to allocate a different carrier frequency to different transmitters or work on a single carrier frequency. The use of a single frequency simplifies the receiver (only one receiving frequency channel is needed), but makes the detection procedure more complex. It also complicates the direct signal and clutter cancellation procedure. The single frequency net transmitters can emit independent noise signals or use the same noise signal in each transmitter. In this case it is necessary to synchronize all the transmitters (e.g., using GPS technology) and transmit the noise signal to all transmitting and receiving nodes. In a synchronous noise radar network we may simply cancel the direct signal as well as the ground clutter using the procedure described in Chapter 4, but each target will generate multiple echoes which must be combined with tracks during the target tracking procedure.

The use of multistatic configuration will increase the accuracy of target positioning. The measurement error area for each transmitter-receiver pair forms a long ellipsoid. Having several independent measurements originating from different transmitter-receiver pairs, it is possible to obtain a much higher pointing accuracy, as depicted in Figure 5.15. The use of two transmitter-receiver pairs allows us to precisely localize the target in two-dimensional space, while the use of three transmitter-receiver pairs allows for three-dimensional pointing and tracking [8, 14–16]. A higher number of transmitter-receiver pairs will increase the pointing accuracy [15].

Multistatic operation increases the likelihood of target detection and decreases the blind zone. In Figure 5.16 the bistatic coverage for a multistatic noise radar with three transmitters and one receiver is presented. The coverage

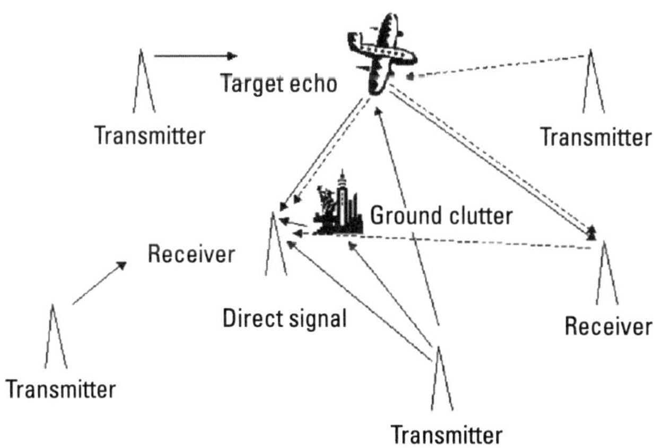

Figure 5.14 Multistatic net of noise radars.

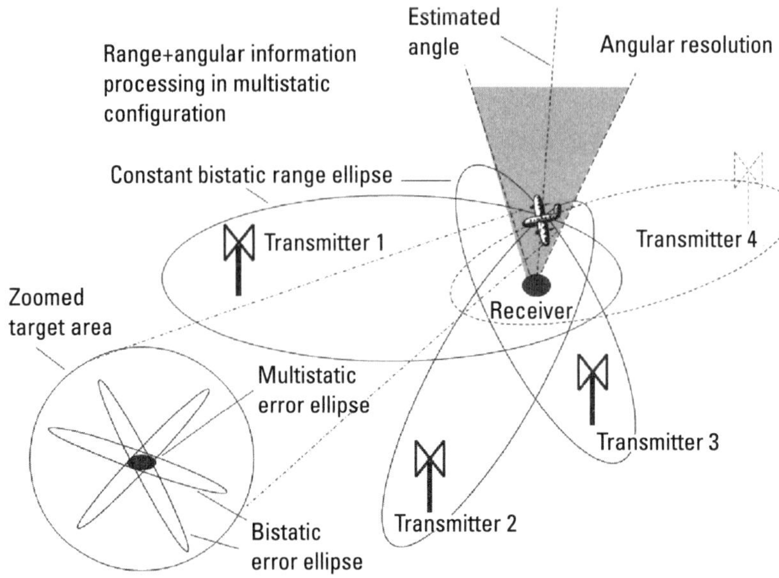

Figure 5.15 Target pointing in a multistatic net of noise radars.

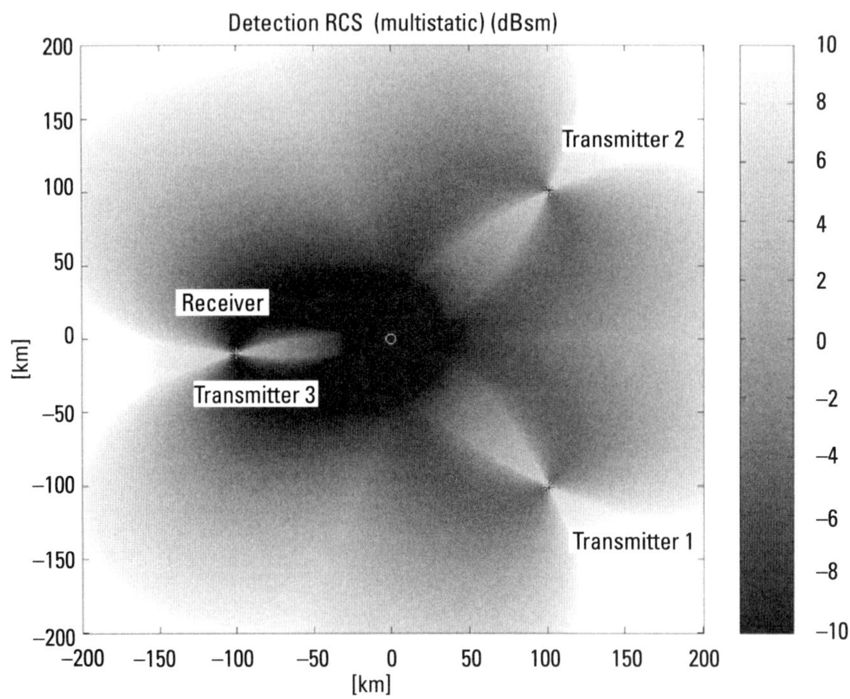

Figure 5.16 Coverage of a multistatic net of noise radars consisting of one receiver and three transmitters. Grayscale: detected RCS (dBsm).

in this figure is defined as the area where the target with a given radar cross section can be detected. As a result, multistatic radar coverage defined in this way is the logical (or rather geometrical) sum of the coverage of each transmitter-receiver pair.

Precision two-dimensional tracking can be performed in the area where a target is detected using signals from at least two transmitters. The precision two-dimensional tracking area presented in Figure 5.17 is smaller than the detection area presented in Figure 5.16, but there are no blind zones between the receiver and the transmitters. For precision three-dimensional tracking, it is necessary to have simultaneous detection which has originated from at least three transmitter-receiver pairs, so the three-dimensional tracking area is much smaller (see Figure 5.18), and blind zones towards the transmitters exist. The three-dimensional tracking capability exists only when transmitters, receivers, and targets are at different Z (high) coordinates. When they all are on this same plane, the effect of geometrical dilution of precision (GDOP) exists and accuracy of Z-coordinate is very low. In general, altitude errors are several times higher than the errors in the horizontal coordinates and depend on the altitude of the target. Z-coordinate accuracy of low-flying targets is usually low.

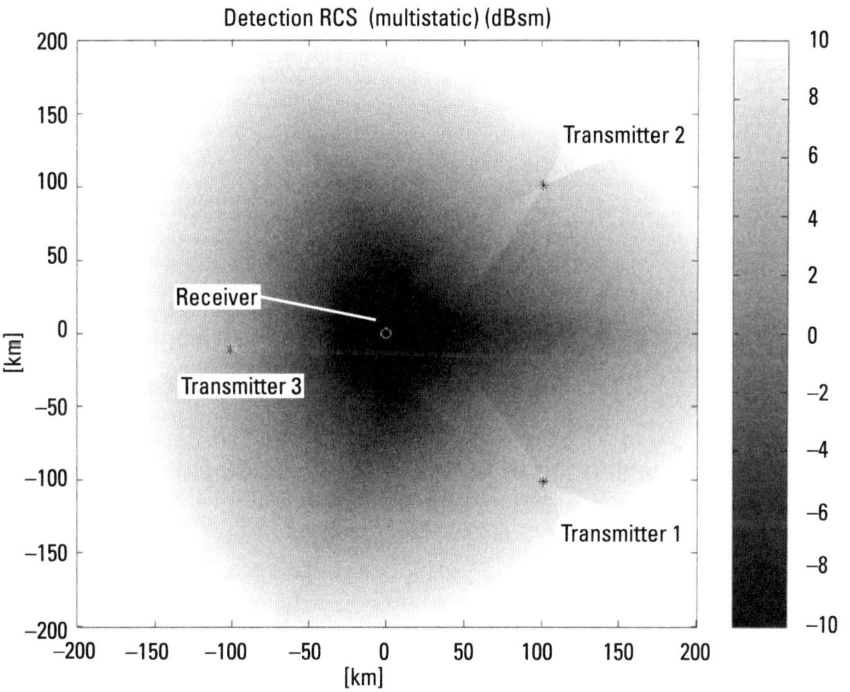

Figure 5.17 Precision two-dimensional tracking area for a multistatic net of noise radars consisting of one receiver and three transmitters. Grayscale: tracked RCS (dBsm).

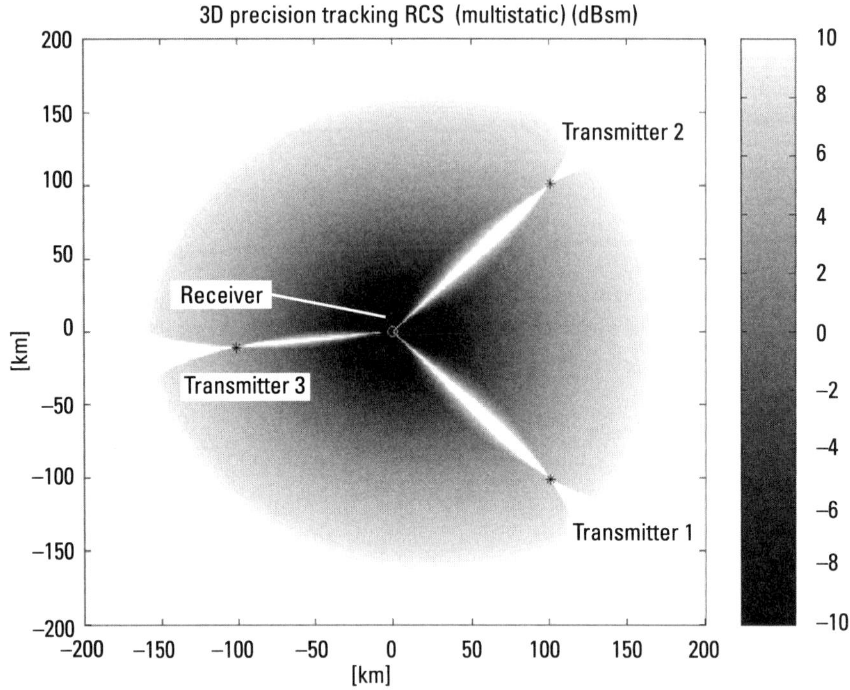

Figure 5.18 Precision three-dimensional tracking area for a multistatic net of noise radars consisting of one receiver and three transmitters. Grayscale: tracked RCS (dBsm).

Multistatic coverage can be enlarged by adding an additional receiving node. The detection coverage for a multistatic net consisting of two receivers and three transmitters is presented in Figure 5.19. In Figure 5.20 the precision two-dimensional tracking area is presented, and in Figure 5.21 the precision three-dimensional tracking area is shown. In the presented configuration each target can be detected by six transmitter-receiver pairs, although only three are needed for precision tracking, so there are no blind zones towards receivers even for the precision three-dimensional tracking.

5.3 Clutter Cancellation in Multistatic and MIMO Noise Radar

As was discussed in Chapter 4, one of the crucial problems of noise radars is the presence of the masking effect [17–21]. Weak targets echoes can be completely masked by sidelobes of strong clutter. This problem can be solved by using adaptive methods for strong echo cancellation [20, 22]. Current research on the noise radar is focused on three main configurations: monostatic, multistatic (netted), and multiple input multiple output (MIMO). The concept of netted noise radars forming a "fence" along a border has been proposed in [23] and the

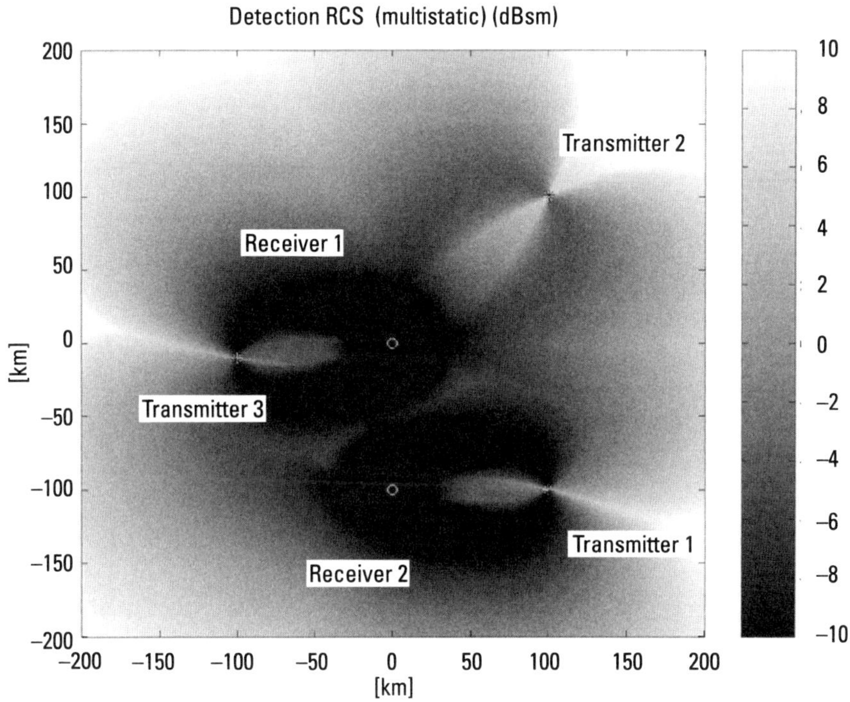

Figure 5.19 Coverage of a multistatic net of noise radars consisting of two receivers and three transmitters. Grayscale: detected RCS (dBsm).

noise MIMO radar has been described in [24]. In the case in which all transmitters working in the multistatic configuration are emitting signals using different carrier frequencies, and the transmitted bands are separated in frequency, then there is no interference between transmitters and the entire signal processing can carried on in the same manner as in the monostatic case described in previous chapters. However, there are many reasons why it would be practical to transmit all signals in the same band. The first reason is the frequency allocation problem and limited frequency resources. It is not trivial to allocate separate frequency bands for all transmitters, especially when the net consists of many of them and uses wideband signals (such as 100 MHz or more). The second is the need of many receiving channels. In a separate frequency multistatic radar network, it is necessary to add to each reviving antenna the number of receivers equal to the number of transmitters. In the case of multiuse of the selected single band, all transmitters use the same (or very similar) carrier frequency using different or the same noise modulation. The use of the same noise modulation leads to a concept very similar to a single-frequency network used in commercial radio or TV broadcasting (DAB, DVB-T). All transmitters must be synchronized (e.g.,

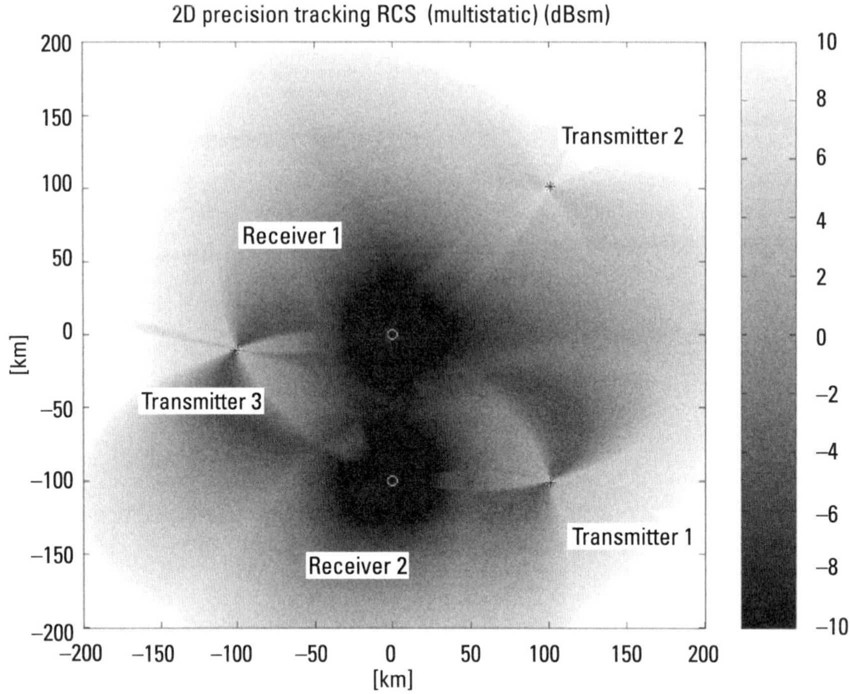

Figure 5.20 Precision two-dimensional tracking area for a multistatic net of noise radars consisting of two receivers and three transmitters. Grayscale: tracked RCS (dBsm).

by GPS or cable synchronization) and transmit the signal of the same carrier frequency and the same modulation.

Although all transmissions have the same contents and differ only in time delay, all methods for signal processing described in Chapters 3 and 4 can be used.

An alternative solution is to transmit independent noise waveforms from each transmitter using the same frequency band. In theory the transmitted noise signals are orthogonal to each other, and due to this orthogonality between components, the signal processing can be relatively simple. The orthogonal components can be removed consecutively (one after another). However, due to limited integration time all transmitted signals are not fully orthogonal and an additional effort has to be made to remove the unwanted signal components to achieve long-range detection of the weak targets.

Let us consider the multistatic noise radar net consisting of K transmitters and L receivers, located in different positions. The transmitters send independent noise signals $x_k(T)$ in the same frequency band. Each of the receivers, apart from the echo signal, can receive (e.g., using dedicated data links or highly directive antennas) the reference signals $x_1(T) \ldots x_K(T)$ sent by all transmitters.

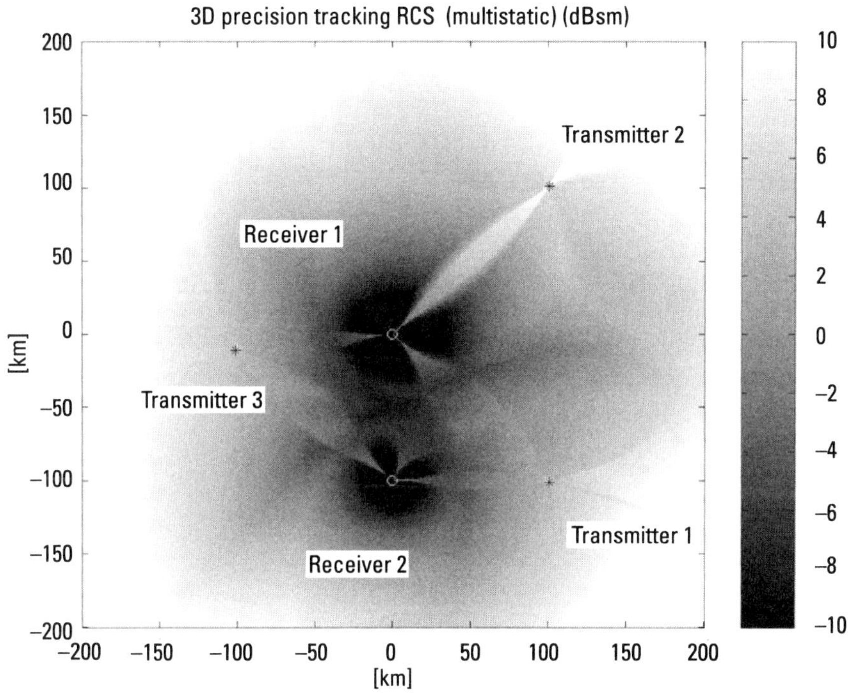

Figure 5.21 Precision three-dimensional tracking area for a multistatic net of noise radars consisting of two receivers and three transmitters. Grayscale: tracked RCS (dBsm).

An example radar net with three transmitters ($K = 3$) and one receiver ($L = 1$) is presented in Figure 5.22.

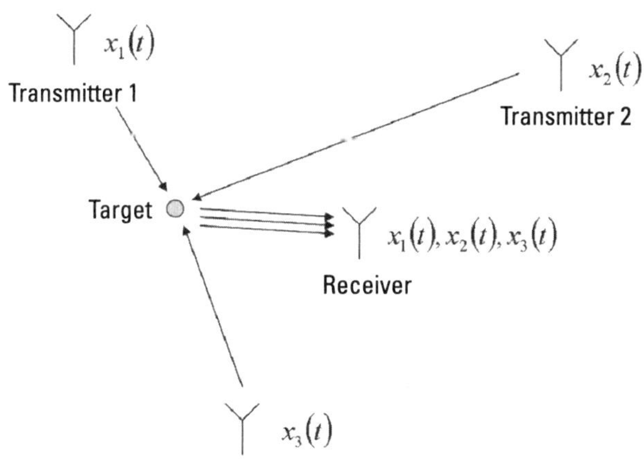

Figure 5.22 The multistatic network with three transmitters working in the same bandwidth and one receiver.

The signal received by the lth receiver originating from the kth transmitter can be expressed by the following formula:

$$y_{l,k}(t) = \sum_{q=1}^{Q} a_q x_k\left(t - \frac{R_q}{c}\right) + \sum_{p=1}^{P} b_p x_k\left(t - \frac{R_p}{c}\right)\exp\left(j2\pi\frac{V_p}{\lambda}t\right) \quad (5.12)$$

where $x_k(T)$ is the signal transmitted by the kth transmitter, aq, bp are the complex signal amplitudes of the stationary and moving targets, respectively, R_q, R_p are the bistatic ranges to the stationary and moving targets, V_p is the target bistatic velocity of the moving targets, λ is the wavelength, Q is the number of stationary targets, and P is the number of moving targets.

The first term on the right side of (5.12) corresponds to the reflections from stationary targets, treated as the ground clutter. The second term represents contributions from moving targets of interest that have to be detected and tracked by the radar net. Although there are K transmitters, the signal received by the lth receiver is the sum of the components which have originated from each transmitter and receiver's thermal noise $w(T)$, as described by the following formula:

$$y_l(t) = \sum_{k=1}^{K} y_{l,k}(t) + w(t) \quad (5.13)$$

Detection of the useful targets is based on the calculation of the range-Doppler correlation function. However, in this case there are K reference signals, so the range-Doppler correlation is calculated for each pair of transmitter-receivers:

$$\psi_{k,l}(R,V) = \int_t y_l(t)\, x_k^*\left(t - \frac{R}{c}\right)\exp\left(-j2\pi\frac{V}{\lambda}t\right)dt \quad (5.14)$$

The signal $x_k(T)$ correlates only with the corresponding term $y_{k,l}(T)$ of the signal $y_l(T)$. However, other components of signals $y_l(T)$ originating from other transmitters act as additional noise sources. As a result, the noise floor of the range-Doppler surface is raised up and this leads to detection losses. In the case of heavy noise clutter, moving targets can be completely masked by the sidelobes of the clutter echoes.

Single-channel ground clutter can be removed using the algorithms described in Chapter 4, but those algorithms have been developed under the assumption that there is a single source of illumination. However, it is relatively simple to extend the CLEAN algorithm based on a lattice filter structure (as presented in Figure 5.23) to the vector form.

As in the previous case, the lattice structure is used for clutter space orthogonalization and the lower part of the filter is used for the removal of orthogonal clutter components from the received signal.

The multichannel lattice prediction algorithm can be described by formulas (5.15) through (5.25) [25]. The forward prediction error $\mathbf{f}_m(n)$ is $K \times 1$ vectors is calculated as:

$$\mathbf{f}_{m+1}(n) = \mathbf{f}_m(n) - \Gamma^f_{m+1}\mathbf{b}_m(n-1) \tag{5.15}$$

The backward prediction errors (based on vectors of an orthogonalized version of clutter subspace) are calculated using a similar formula:

$$\mathbf{b}_{m+1}(n) = \mathbf{b}_m(n-1) - \Gamma^b_{m+1}\mathbf{f}_m(n) \tag{5.16}$$

The $(K \times K)$ matrices of partial correlation "PARCOR" coefficients $\Gamma^f_{m+1}, \Gamma^b_{m+1}$ are calculated as:

$$\Gamma^f_{m+1} = E\left[\mathbf{f}_m(n)\mathbf{b}^H_m(n-1)\right]\left[\mathbf{P}^b_m\right]^{-1} \tag{5.17}$$

$$\Gamma^b_{m+1} = E\left[\mathbf{b}_m(n-1)\mathbf{f}^H_m(n)\right]\left[\mathbf{P}^f_m\right]^{-1} \tag{5.18}$$

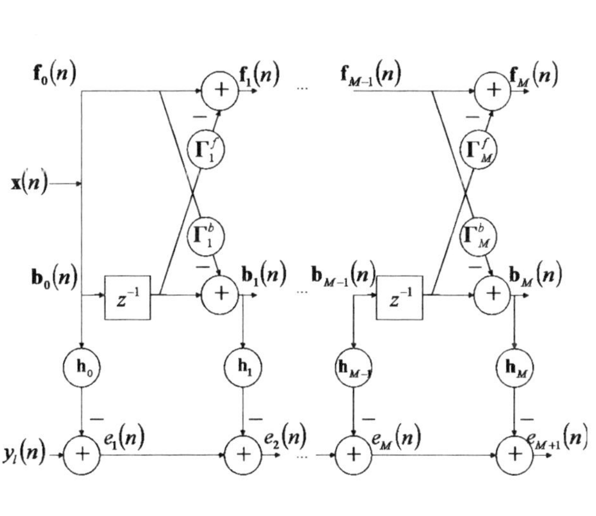

Figure 5.23 The lattice filter structure to remove a ground clutter component from the received signal.

and the $(K \times K)$ covariance matrices $\mathbf{P}_m^f, \mathbf{P}_m^b$ of the forward and backward prediction errors are defined as follows:

$$\mathbf{P}_{m+1}^f = \left[\mathbf{I} - \boldsymbol{\Gamma}_{m+1}^f \boldsymbol{\Gamma}_{m+1}^b\right]\mathbf{P}_m^f \tag{5.19}$$

$$\mathbf{P}_{m+1}^b = \left[\mathbf{I} - \boldsymbol{\Gamma}_{m+1}^f \boldsymbol{\Gamma}_{m+1}^b\right]\mathbf{P}_m^b \tag{5.20}$$

The recursions (5.15) through (5.20) are repeated M times with initial conditions described as:

$$\mathbf{f}_0(n) = \mathbf{b}_0(n) = \mathbf{x}(n) = \left[x_1(n) \quad x_2(n) \quad \cdots \quad x_K(n)\right]^T \tag{5.21}$$

$$\boldsymbol{\Gamma}_1^f = \boldsymbol{\Gamma}_1^b = \mathbf{R}_{xx}^{-1}(0)\mathbf{R}_{xx}(1) \tag{5.22}$$

$$\mathbf{P}_1^f = \mathbf{P}_1^b = \left(\mathbf{I} - \boldsymbol{\Gamma}_1^f \boldsymbol{\Gamma}_1^b\right)\mathbf{R}_{xx}(0) \tag{5.23}$$

where $x_k(n)$ represents samples of the transmitted signals and $\mathbf{R}_{xx}(m) = E\left[\mathbf{x}(n)\mathbf{x}^H(n-m)\right]$ is the $(K \times K)$ autocorrelation matrix of the reference signals.

To remove clutter components from the received signal, it is necessary to calculate the correlation matrix between the input signal at the consecutive stage of the canceler and the clutter vectors:

$$\mathbf{h}_m(n) = \left[\mathbf{P}_m^b\right]^{-1} E\left[\mathbf{b}_m(n)\mathrm{e}_m^H(n)\right] \tag{5.24}$$

and then remove unwanted clutter components as stated here:

$$\mathrm{e}_{m+1}(n) = \mathrm{e}_m(n) - \mathbf{h}_m^H \mathbf{b}_m(n) \tag{5.25}$$

For $m = 0$, we have $e_0(n) = y_l(n)$ and $\mathbf{h}_0(n) = [\mathbf{R}_{xx}^{-1}(n)]^{-1} E\left[\mathbf{x}(n)y_l^*(n)\right]$, where $\mathrm{y_l(n)}$ represents the samples of the signal received by the lth receiver. The final estimate of the clutter-free input signal, consisting of nonzero Doppler echoes and receiver noise, is given by $e_{M+1}(n)$. In the above equations, expectation operator E[·] has been used. In practical implementation, instead of the expectation operator, a sample mean (based on the summation operator) is used.

To illustrate the effectiveness of the method presented above for clutter cancellation in a multistatic noise radar net working in a single-frequency band, the simulation result of signal processing is presented. In the simulation the transmitted signal was a band-limited Gaussian noise in the 100-kHz band.

The targets were simulated by adding time-delayed and Doppler shifted versions of the transmitted signal to the received signal. The sampling frequency was equal to 200 kHz and the length of the signal blocks was equal to 32,768. The range-Doppler surfaces were calculated using formula (5.14). To reduce the Doppler sidelobes a Hamming window was applied to the signals before correlation.

The simulated scenario is presented in Figure 5.22. Three transmitters and one receiver were simulated. Ten stationary targets of a different radar cross section and three moving targets were simulated in the presented example. Figure 5.24 presents the result of the correlation of the signal transmitted by the first transmitter $x_1(T)$ with the received signal $y_1(T)$. The values of the correlation were clipped at the mean noise floor level. In the figure, only ground clutter at zero Doppler frequency is visible. The noise floor level is approximately at 30 dB (arbitrary scale). As can be seen in Figure 5.24, no moving targets are visible, as they are masked by the ground clutter echoes that have originated from all transmitters.

The effect of the zero Doppler clutter removal procedure using only the signal from the first transmitter $x_1(T)$ is presented in Figure 5.25. The single-channel lattice filter in this example completely removed the ground echoes that originated from the first transmitter, so no ground clutter echoes are visible.

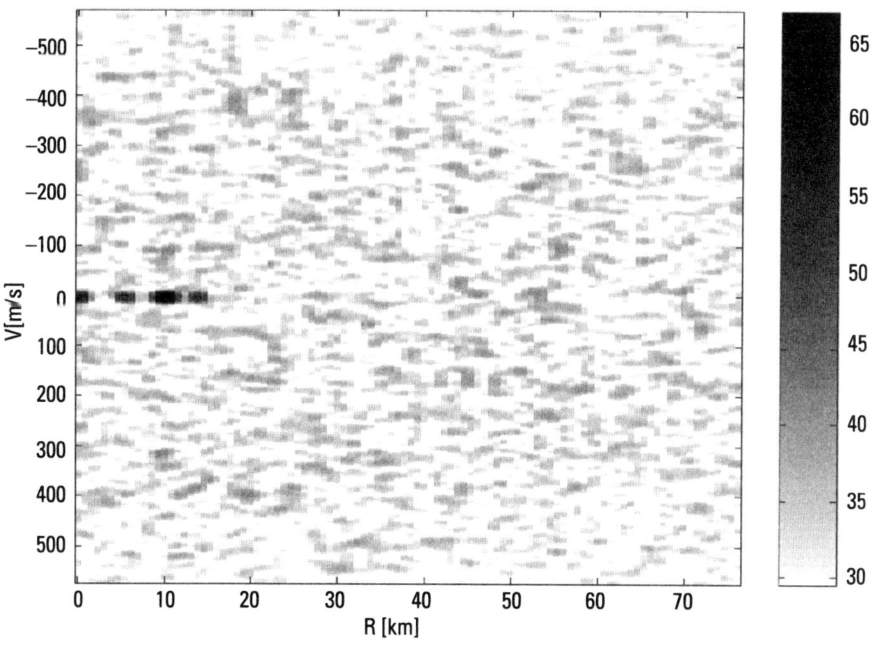

Figure 5.24 Correlation of the signal from the first transmitter $x1(T)$ with the received signal $y1(T)$.

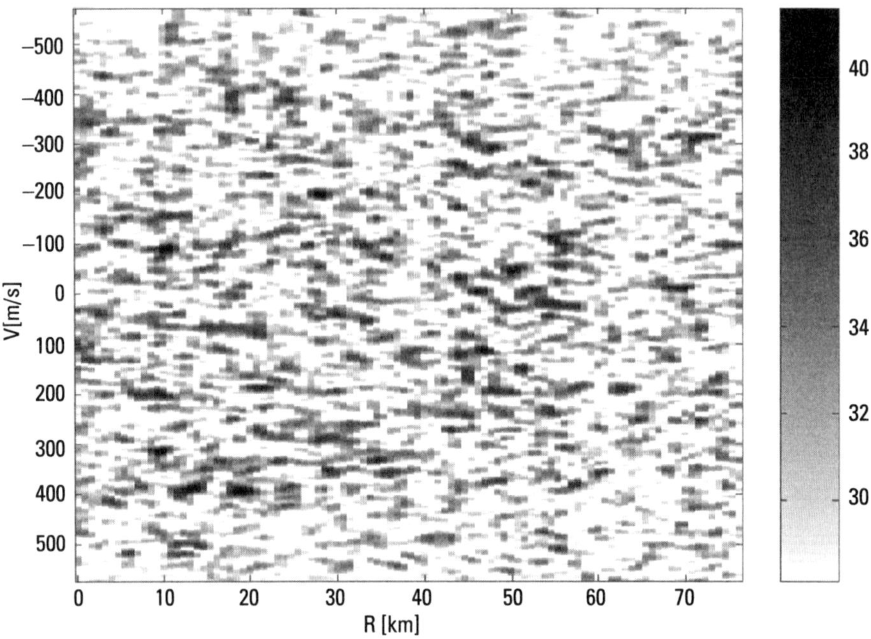

Figure 5.25 Correlation of the signal from the first transmitter $x1(T)$ with the received signal $y1(T)$ after clutter removal from the first channel.

However, the clutter components that originated from the other transmitters mask all moving targets' echoes, and only the noise floor at the level of 28 dB is visible. The first echo cancellation reduces the noise floor by only 2 dB.

In the consecutive steps, a single-channel lattice filter was used to remove signals originating from consecutive transmitters separately. This led to further reduction of the noise floor to the level of –5 dB. This means that the noise floor was reduced in total by 35 dB. The range-Doppler plane calculated after such a clutter cancellation procedure is presented in Figure 5.26. Due to the lack of orthogonality between the transmitted signals, the ground clutter components have not been removed completely, but the 35-dB reduction in the noise floor was sufficient to reveal the moving targets' echoes.

The use of the multichannel lattice filter described by formulas (5.15) through (5.25) removed the ground clutter components completely, as depicted in Figure 5.27. In the presented case the noise floor was reduced by an additional 19 dB to the level of –24 dB. The simulated moving targets are clearly visible above the noise floor.

As shown in Figures 5.24 through 5.27, the use of the single-channel ground clutter cancellation procedure does not CLEAN the received signal sufficiently, while vector CLEANing provides satisfying results. In Figure 5.28 the noise floor level versus the number of transmitters is depicted for different

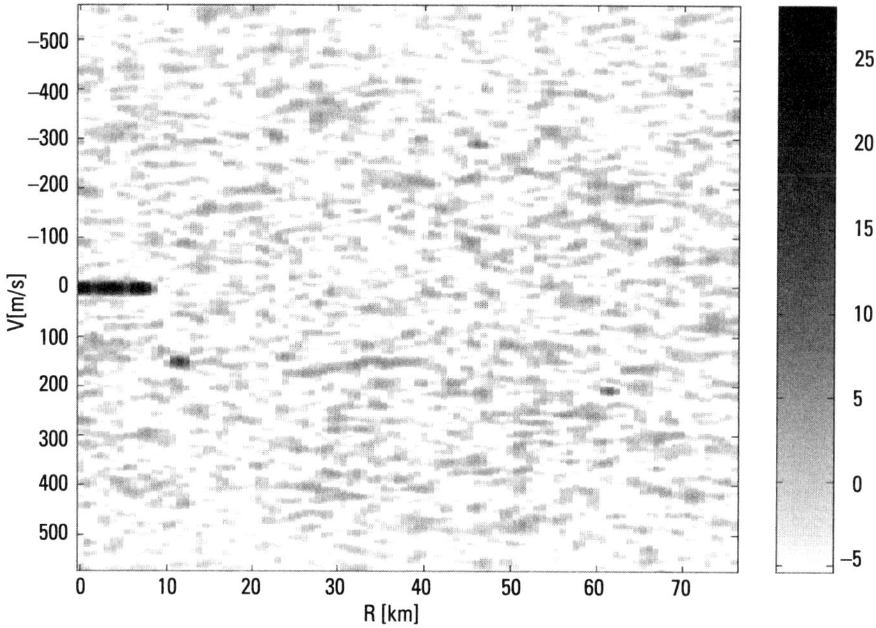

Figure 5.26 Correlation of the signal from the first transmitter $x1(T)$ with the received signal $y1(T)$ after independent clutter removal from each channel.

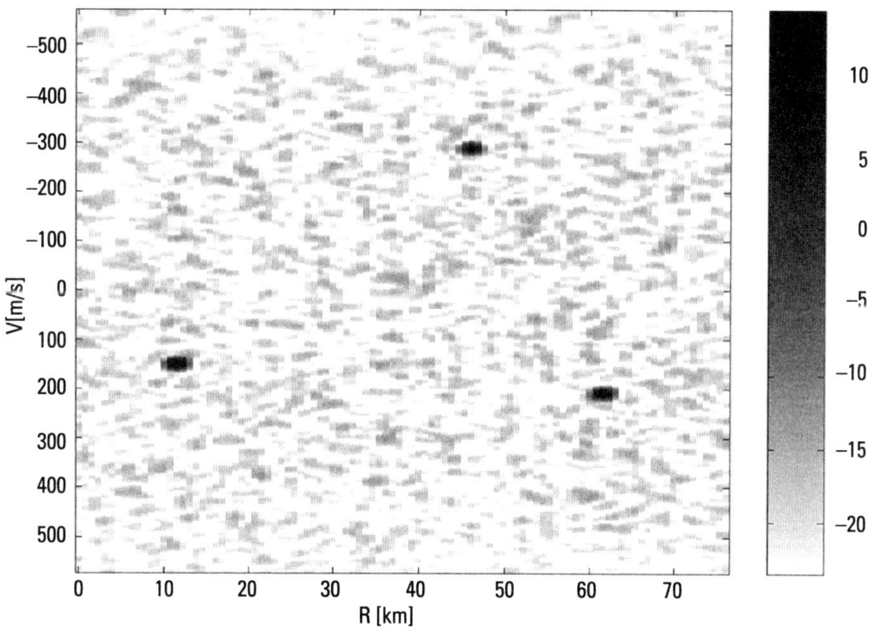

Figure 5.27 Correlation of the signal from the first transmitter $x1(T)$ with the received signal $y1(T)$ after joint clutter removal from each channel by the multichannel lattice filter.

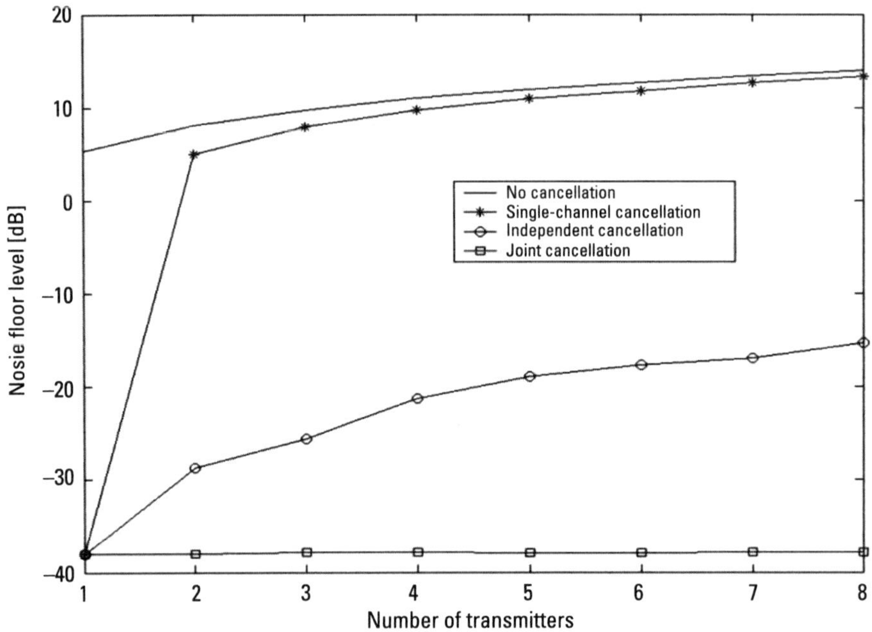

Figure 5.28 Noise floor level for different clutter cancellation methods versus number of transmitters.

clutter cancellation methods. The simulated received signal consists of several strong ground clutter components and the receiver noise is 40 dB below the level of the useful signals. It can be observed that a joint (vector) cancellation method using a multidimensional lattice filter canceled all ground clutter, and the noise floor level is independent of the number of transmitters. This fixed noise floor level results from the noise added to the signal. Independent cancellation of each channel causes the noise floor to rise by several decibels in comparison to the multidimensional lattice method. When the clutter removal method is not used, the noise floor level increases steadily as a result of an increase in the total radiated power (sum of the power radiated by all transmitters) of the larger number of transmitters.

References

[1] Willis, N. J., *Bistatic Radar*, Norwood, MA: Artech House, 1991.

[2] Griffiths, H. D., et al., "Measurement and Analysis of Ambiguity Functions of Off-Air Signals for Passive Coherent Location," *Electronics Letters*, Vol. 39, No. 13, June 26, 2003, pp. 1005–1007.

[3] Gould, D. M., R. S. Orton, and R. J. E. Pollard, "Forward Scatter Radar Detection," *RADAR 2002*, IEE Conf. Publ. No. 490, October 15–17, 2002, pp. 36–40.

[4] Kabakchiev, C., et al., "Forward Scatter Radar Detection and Estimation of Marine Targets," *2012 13th International Radar Symposium (IRS)*, Military University of Technology, Poland, Warsaw, May 23–25, 2012, pp. 533–538.

[5] Gashinova, M., et al., "Signal Detection in Multi-Frequency Forward Scatter Radar," *2010 European Radar Conference (EuRAD)*, September 30–October 1, 2010, pp. 276–279.

[6] Cherniakov, M., et al., "Analysis of Forward Scattering Radar for Ground Target Detection," *European Radar Conference 2005 (EURAD 2005)*, October 6–7, 2005, pp. 145–148.

[7] Sizov, V., M. Cherniakov, and M. Antoniou, "Forward Scattering Radar Power Budget Analysis for Ground Targets," *IET Radar, Sonar & Navigation*, Vol. 1, No. 6, December 2007, pp. 437–446.

[8] Chernyak, V. S., *Fundamentals of Multisite Radar Systems: Multistatic Radars and Multiradar Systems*, Boca Raton, FL: CRC Press, 1998.

[9] Sahr, J. D., and F. D. Lind, "The Manastash Ridge Radar: A Passive Bistatic Radar for Upper Atmospheric Radio Science," *Radio Sci.*, Vol. 32, No. 6, 1997, pp. 2345–2358.

[10] Skolnik, M. I., *Radar Handbook*, 3rd ed., New York: McGraw-Hill, 2008.

[11] Kulpa, K., and Z. Czekała, "Long-Range Performance Increase in Passive PCL Radar," *3rd Multinational Conference on Passive and Covert Radar 2003 (PCR-2003)*, University of Washington Applied Physics Laboratory, Seattle, WA, October 21–23, 2003.

[12] Kulpa, K., "Adaptive Clutter Rejection in Bi-Static CW Radar," *Proc. Int. Radar Symp. IRS-2004*, Warsaw University of Technology, Warsaw, Poland, May 2004, pp. 61–66.

[13] Griffiths, H. D., "From a Different Perspective: Principles, Practice and Potential of Bistatic Radar," *IEEE Proc. International Conference RADAR 2003*, Adelaide, Australia, September 3–5, 2003, pp. 1–7.

[14] Gray, D. A., "Multi-Channel Noise Radar," *International Radar Symposium IRS-2006*, Warsaw University of Technology, Krakow, Poland, May 24–26, 2006, pp. 417–420.

[15] Say, K., *Statistical Error Analysis of a DOA Estimator for a PCL System Using the Cramer-RAO Bound Theorem*, thesis, Air Force Institute of Technology, School of Engineering and Management, Wright-Patterson AFB, OH, 2002.

[16] Malanowski, M., and K. Kulpa, "Two Methods for Target Localization in Multistatic Passive Radar," *IEEE Transactions on Aerospace and Electronic Systems*, Vol. 48, No. 1, January 2012, pp. 572–580.

[17] Kulpa, K., M. Malanowski, and Z. Gajo, "Ground Clutter Cancellation in MIMO and Multistatic Noise Radars," *2007 IET International Conference on Radar Systems*, October 15–18, 2007, pp. 1–5.

[18] Kulpa, K. S., and Z. Czekala, "Masking Effect and Its Removal in PCL Radar," *IEE Proceedings Radar, Sonar and Navigation*, Vol. 152, No. 3, June 3, 2005, pp. 174–178.

[19] Kulpa, K., and Z. Czekala, "Short Distance Clutter Masking Effects in Noise Radars," *J. Applied Radio Electronics*, Vol. 4, No. 1, 2005, pp. 96–98.

[20] Kulpa, K. S., and Z. Czekała, "Ground Clutter Suppression in Noise Radar," *Proc. Int. Conf. RADAR 2004*, Toulouse, France, October 18–22, 2004, p. 236.

[21] Kulpa, K., and Z. Czekała, "Long-Range Performance Increase in Passive PCL Radar," *3rd Multinational Conference on Passive and Covert Radar, 2003 (PCR-2003)*, University of Washington Applied Physics Laboratory, Seattle, WA, October 21–23, 2003.

[22] Axelsson, S. R. J., "Improved Clutter Suppression in Random Noise Radar," *Proc. URSI 2005 Commission F Symposium on Microwave Remote Sensing of the Earth, Oceans, Ice, and Atmosphere*, Barza d'Ispra, Italy, April 20–21, 2005.

[23] Salminen, V. -J., et al., "Noise Modulated Multistatic Surveillance Radar Concept," *International Radar Symposium 2006 (IRS 2006)*, Warsaw University of Technology, Poland, May 24–26, 2006, pp. 1–4.

[24] Gray, D. A., "Multi-Channel Noise Radar," *International Radar Symposium 2006 (IRS 2006)*, Warsaw University of Technology, Poland, May 24–26, 2006, pp.1–4.

[25] Marple, S. L., *Digital Spectral Analysis*, Upper Saddle River, NJ: Prentice-Hall, 1987.

6

Noise Synthetic Aperture Radar

The first application of the classical radar was to detect ships and airplanes. After a few years of development, it was found that radar could also be used for imaging purposes. One of the first imaging applications was bombsight, a device used in bomber aircraft to help in finding the optimal time instant to drop bombs. Further developments led to coherent imaging techniques: Doppler beam sharpening (DBS) [1–4] and synthetic aperture radar (SAR) [5–8]. For a long time detailed pictures of the Earth's surface, taken from airplanes and satellites, have been used for monitoring natural terrestrial phenomena and human activities. In civil applications these pictures are used for map creation, biomass calculation, flood control, pollution control, and others. In military applications detailed pictures are used most frequently for intelligence, the control of army movement, target recognition, identification of military installations, and others. Pictures taken in the visible light spectrum, as well as in near infrared or ultraviolet, are susceptible to atmospheric conditions and, of course, to the Sun's illumination. The search for sensors that are able to operate day and night in all weather conditions has led to the development of airborne and spaceborne SAR technology. This technology makes it possible to produce medium- and high-resolution images of an area of interest. The principle of SAR technology is described in detail in [5–8]. Therefore, in this chapter the SAR principles are described only briefly.

The range resolution of a conventional radar depends on the transmitted signal bandwidth, and can reach values from 1 cm to 100m. The cross-range resolution depends on the width of the antenna main beam, and is usually much worse (20 to 1,000m at a 10-km distance). To increase the cross-range resolution of the classical radar, it is necessary to build extremely large (long) antennas. Instead of constructing a very long physical antenna, it is possible to process the signals collected during the flight over a selected path and form a very long synthetic antenna aperture. The spatial resolution of an SA image depends on the bandwidth of the transmitted signal (in the range dimension) and the size of the antenna aperture (in the cross-range dimension). The theoretical cross-range resolution limit of SAR is equal to half of the real antenna aperture length and is independent of the range to the target. To increase the image resolution, it is necessary to decrease the real antenna size; however, the use of small antennas leads to several technical problems. The transmitted signal power must be high enough to guarantee the detection and proper imaging of small targets at long distances. Due to a wide beam of the real antenna, the Doppler frequency spectrum of the ground returns is also wide (especially in the case of high platform velocity) and high pulse repetition frequencies are required. This effect decreases the maximum unambiguous range and thus limits the width of the observation strip.

The designers of SAR systems are looking for new technologies and processing algorithms that could overcome limitations inherent to classical SAR technology and lead to advanced SAR radars with better spatial resolution and wider observed areas. One of the technologies that can be used for this purpose is noise radar technology. In Section 6.2 the algorithm of noise SAR processing is presented, and in Section 6.3 the results of the image enhancement algorithm are shown.

6.1 Fundamentals of Classical SAR Processing

For a long time engineers looked for a method of obtaining high-quality images using radars installed on board an airplane. The classical surveillance radar with a short sounding pulse gave a good range resolution (usually 10–100m), while the cross-range (azimuth) resolution was much worse and depended on the range from the radar to the imaged area. For narrow-beam radars cross-range resolution is a product of range and antenna beamwidth. For a 20-mrad (1.1°) antenna beam, the cross-range resolution at a distance of 10 km is equal to 200m, and for the distance 100 km it is 2 km. The increase of cross-range resolution in the classical radar may be achieved by decreasing the antenna beamwidth. The theoretical width of an antenna beam is approximately equal to the wavelength divided by antenna length (λ/La), so to improve the cross-range resolution, it is possible to use either a longer antenna (in the along-track dimension) or shorter waves. The usage of very short waves is limited by the atmospheric attenuation effect, especially in the presence of high humidity, fog, or clouds. Because of this limitation the L-, C-, S-, X-, and K-bands are used in airborne radars. The length of the radar antenna is limited by the size of the airplane. In the case of surveillance radars the antenna length is limited by the width of the airplane. In a small airplane it is possible to install a rotating antenna with length of between 0.5 to 2m. For example, a 1.5m antenna working in the X-band has a beamwidth of 20 mrad. As stated before, the cross-range resolution of a radar equipped with such an antenna is moderate. A longer antenna may be installed only as a fixed antenna on the side of an airplane. In such a case the length of an airplane limits the antenna length. Additional antenna length limitations may have their origin in antenna vibration and thermal deformation. The typical antenna length used in side-looking antenna radars is 4 to 10m. An X-band 10-m antenna has a beamwidth of 3 mrad, so the cross-range resolution at a 10km range is equal to 30m and is at the range of 100 km to 300m. However, using the side-looking antenna radar, it is possible to create a radar image of terrain on the side of the plane trajectory. The image is formatted when the plane passes by the imaged area.

Research on a completely new technique for high-resolution radar imaging was carried out in the 1950s and 1960s [9–11]. In 1965, an American scientist named Carl Wiley patented the new concept of the application of Doppler information to increase radar resolution [12]. Since that time, this concept has become known as the precursor of the new field of radar technology, the synthetic aperture radar [13, 14].

The idea of SAR technology is based on the coherent summation of return echo signals collected along the straight flight of the airplane. The coherent summation of a signal is equivalent to forming a very long "synthetic" antenna with a narrow beam. A typical SAR scenario is presented in Figure 6.1.

Let us assume that a synthetic aperture radar installed on a moving platform (airplane or satellite) is equipped with an antenna of real length L_a (in the along-track dimension) and a height of W_a. The main beam of the antenna is

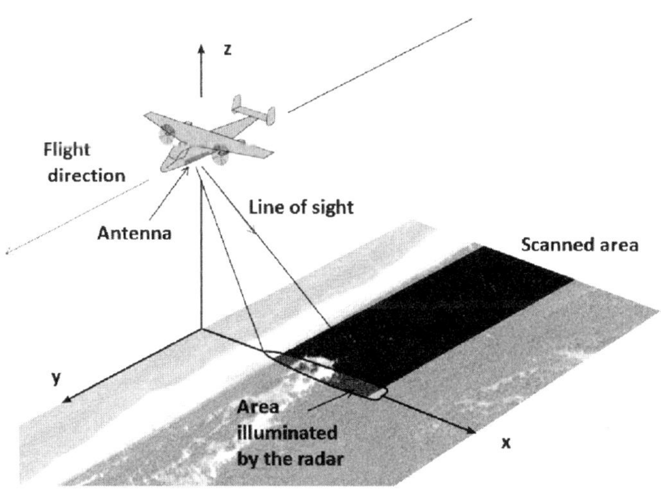

Figure 6.1 Typical SAR scenario.

perpendicular to the flight-line and directed at the looking angle γ towards the Earth's surface, as depicted in Figures 6.2 and 6.3.

The 3-dB along-track beamwidth of the antenna is $\theta_H \approx \lambda / L_a$ (assuming rectangular antenna illumination), and the cross-track beamwidth is $\theta_v \approx \lambda / W_a$. The width of the scanning strip γ (assuming a narrow-beam antenna and a relatively small-looking angle W_g) can then be evaluated as:

$$W_g \approx \frac{\lambda R_m}{W_a \cos\gamma} = \frac{\lambda H}{W_a \cos^2\gamma} \tag{6.1}$$

The classical pulse SAR radar emits sounding radio pulses of duration τ and the carrier frequency F. The radar slant-range resolution is equal to $c\tau/2$, and ground range resolution (in the middle of the beam) is equal to:

$$\Delta R_g \approx \frac{\Delta R_s}{\sin\gamma} = \frac{c\tau}{2\sin\gamma} \tag{6.2}$$

The ground range resolution is not constant and changes reciprocally with the sine of the angle between the line of sight and the normal to the Earth's surface. For typical look angles (30°–60°) the ground resolution is 1.5 to 2 times worse than the slant range resolution of the radar. For an angle close to

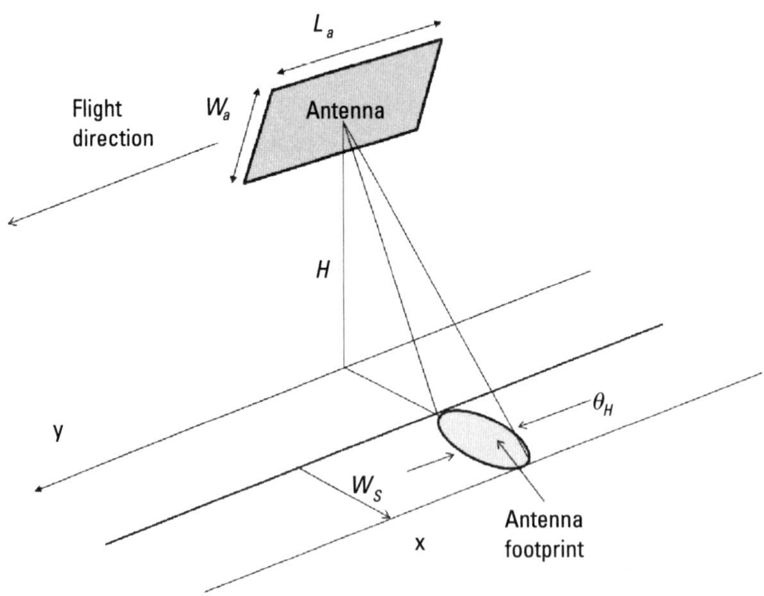

Figure 6.2 Geometry of the SAR system.

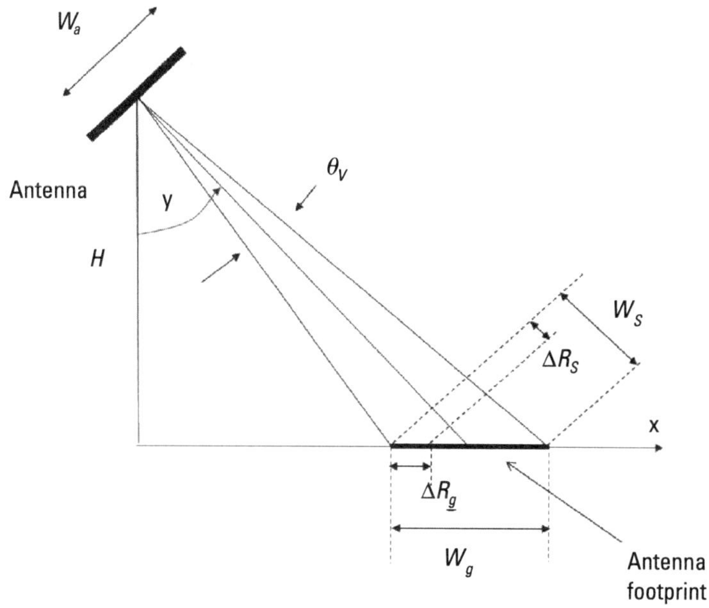

Figure 6.3 SAR resolution in the cross-track direction.

90°, the range resolution is almost equal to the slant range resolution. However, such large angles are not of interest in practice because of a very heavy shadow effect, as presented in Figure 6.4.

This effect is very clearly visible in urban and mountain regions. Objects located in the shadowed region are invisible in the final SAR image. In some

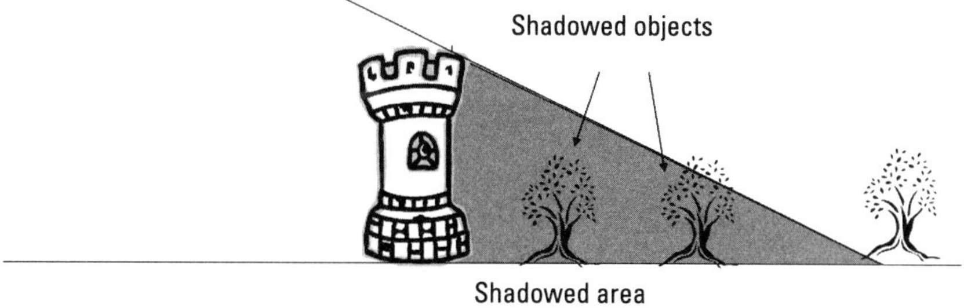

Figure 6.4 SAR shadow effect.

cases this effect can be useful, while it is possible to estimate the height of the objects by measuring the length of the SAR shadow.

The cross-range (along the track) illumination area diameter can be predicted by the formula:

$$W_H \approx \frac{\lambda R}{L_a} = \frac{\lambda H}{L_a \cos\gamma} \tag{6.3}$$

For surveillance or side-looking airborne radar (SLAR), radars the diameter W_H (6.3) is equal to the radar cross-range resolution. The cross-range resolution can be significantly improved by applying signal processing techniques to process the received signal further.

The first step for deriving an appropriate processing schema is to predict the echo signal which originated from a single scattering point. Let us consider a single scatterer placed on the Earth's surface at the coordinates (X_o, 0,0), and a SAR airplane traveling along the Y-axis at the height H with velocity v. The range between the radar and the scattering point is equal to

$$r(t) = \sqrt{H^2 + X_0^2 + (v \cdot t)^2} = \sqrt{R^2 + (v \cdot t)^2} \tag{6.4}$$

This range can be presented in the equivalent Taylor series form:

$$r(t) = R + \frac{(v \cdot t)^2}{2R} + \ldots \tag{6.5}$$

For narrow antenna beams the higher terms may be neglected, while for wider antenna beams it is necessary to use the exact formula (6.4) or add the higher-order terms to (6.5). The received echo signal, downconverted to baseband, can be modeled by the equation:

$$x_r(t) = A(t)\exp(j\varphi(t)) \tag{6.6}$$

where $A(t)$ denotes the return signal (echo) amplitude depending on the transmitted signal power, the antenna radiation pattern, the radar platform velocity, and the range to scatterer and its cross-section, and $\phi(t)$ denotes the received signal phase equal to:

$$\phi(t) = \phi_o - 2 \cdot \frac{2\pi \cdot r(t)}{\lambda} \tag{6.7}$$

The relation (6.6) is valid for the time interval: $t \in \left\langle -\frac{T_{ob}}{2}, \frac{T_{ob}}{2} \right\rangle$, where T_{ob} is the scatterer observation time:

$$T_{ob} = \frac{W_H}{v} = \frac{\theta_H \cdot R}{v} = \frac{\lambda R}{L_a v} \tag{6.8}$$

Relation (6.8) can also be used for a wider time interval by setting $A(t)$ = 0 outside interval $\left\langle -\frac{T_{ob}}{2}, \frac{T_{ob}}{2} \right\rangle$.

Substituting the target range (6.5) into (6.7), one can obtain the received signal phase in the simplified form:

$$\phi(t) = \phi_o - \frac{4\pi}{\lambda}\left[R + \frac{(v \cdot t)^2}{2R} + \ldots \right] \tag{6.9}$$

The received signal Doppler frequency (after neglecting the higher terms of the Taylor series) is equal to:

$$f(t) = \frac{1}{2\pi}\frac{d\phi(t)}{dt} \approx -\frac{2v^2}{\lambda R}t \tag{6.10}$$

It can be easily seen that the signal frequency is a linear function of time, so the signal (6.6) is a classical linear frequency modulated (LFM) chirp signal. Combining (6.10) and (6.8), one can obtain the Doppler bandwidth of the point scatterer signal, equal to:

$$B = \frac{2v}{L_a} \tag{6.11}$$

As the received signal is an LFM, it is possible to apply the classical pulse compression technique described in Chapter 2. The focused image will be produced by the matched filtering of the received signal in the cross-range dimension (2.13). It is worth noting that the pulse response of the compression filter is different for different ranges, while the observation time (6.8) depends linearly on the range to the scattering center and the bandwidth (6.11) is range independent.

The distance between the radar and the point scatterer is a function of time (see (6.4)). The change of that distance produces not only the phase shift of the received signal, but also the target range migration effect. If the target range migration, described by the formula:

$$\Delta R(t) = \sqrt{R^2 + (v \cdot t)^2} - R \tag{6.12}$$

is smaller than half of the radar range resolution ($\Delta R(t) < c\tau/2$), then the target echo will remain in the same range gate during the whole observation time. In this case the matched filter can be designed as a single-dimension filter working in the along-track dimension. The maximum range for which the single-dimension compressing filter can be used for image formation is limited by the following inequality:

$$R < \frac{2L_a^2 c\tau}{\lambda^2} \tag{6.13}$$

If the range to the target does not fulfill the constraint (6.13), it is necessary to perform the two-dimensional range-cross-range filtration. The filtration in the time domain (or the equivalent spatial one) requires great computational power, and a number of much more effective algorithms in the frequency domain have been developed [8].

The single point scatterer echo after pulse compression (performed either in the time or frequency domains) will have (in the time dimension related to the cross-range spatial dimension) a pulse width equal to the reciprocal of the signal Doppler frequency bandwidth. Thus, the cross-range resolution ΔR_{SAR} of the focused SAR is equal to the product of the platform speed and the reciprocal of the signal Doppler bandwidth. After a simple calculation, one can obtain the final SAR resolution formula [6]:

$$\Delta R_{SAR} = \frac{v}{B} = \frac{L_a}{2} \tag{6.14}$$

As stated in (6.14) the cross-range SAR resolution is independent of the range to the target and the carrier frequency (wavelength). It depends only on the SAR real antenna aperture length. A better resolution, opposed to the surveillance radar, can be obtained using a smaller antenna. The decrease of the antenna length improves the SAR system cross-range resolution. However, the reduction of the SAR antenna size also has some negative effects. It leads to the decrease of radar sensitivity, as the antenna gain depends on the antenna size. It also increases the required pulse repetition frequency and, as a result, decreases the width of the scanned strip.

An example of a focused SAR image created by the author's team is presented in Figure 6.5 [15]. The image was created using off-line processing. The raw radar data was recorded onboard a moving platform equipped with a radar working in the X-band. The flight altitude was 500m and the platform speed was 70 m/s. The mean distance between the airplane and the imaged area was 20 km. The available navigation data (true heading, ground speed) was not precise enough, so it was necessary to apply an autofocusing algorithm [16–21]. The range resolution of the image is 15m, while the cross-range resolution is equal to 0.6m and the antenna size was 1.3m.

The SAR radar detection range can be predicted using (2.8). In the focused SAR the integration time is equal to the scatterer observation time described by (6.8). This time depends on the range. As a result, the energy effectively used for scene illumination is equal to the product of the mean emitting power P_{Tm} and observation time, and depends on the range:

$$E_T = P_{Tm} T_{ob} = P_{Tm} \frac{\lambda R}{L_a v} \tag{6.15}$$

Substituting (6.15) into (2.8), one can obtain the SAR detection range:

Figure 6.5 Focused SAR image, Polish coast near the town of Leba.

$$R_{\max} = \sqrt[3]{\frac{P_{Tm} G_T G_R \lambda^3 S_o}{(4\pi)^3 LkT_R D_o L_a v}} \tag{6.16}$$

Let us assume that the SAR is equipped with a rectangular antenna of the size ($L_a x W_a$) with the (theoretical) gain $4\pi L_a W_a / \lambda^2$. In such a case the SAR range equation (6.16) can be rewritten in the form:

$$R_{\max} = \sqrt[3]{\frac{P_{Tm} W_a^2 L_a S_o}{(4\pi) Lk T_R D_o \lambda v}} \tag{6.17}$$

As follows from (6.17), the SAR detection range is proportional to the cube root of the transmitted signal power (rather than the fourth root for a classical surveillance radar), the cube root of the antenna length (in the along-track direction), and the cube root of the square of the antenna width (in the cross-track direction) and is inversely proportional to the cube root of the platform speed and wavelength. Because the integration time in the SAR radar is usually much longer than it is for a surveillance radar, the detection range in the SAR mode is usually greater than the detection range in surveillance mode for this same radar. The time-on-target in a surveillance radar is typically at the level of tens of milliseconds. The integration time in a SAR mode, depending on the range, antenna beamwidth and the platform speed, is in the range of seconds or tens of seconds. As a result, in most radars the integration time in the SAR mode time is 10^2 to 10^4 times longer than in the surveillance mode, and thus the detection range in the SAR mode is 3 to 10 times greater than in the surveillance one.

To avoid the distortion of the SAR image caused by the aliasing effect, the pulse repetition frequency *PRF* must be higher than the received signal Doppler bandwidth:

$$PRF \geq \frac{2v}{L_a} \tag{6.18}$$

In classical radars the pulse repetition frequency is limited by the maximum target range $R_{\max}$ and has to fulfill the following inequality:

$$PRF < \frac{c}{2R_{\max}} \tag{6.19}$$

In the scenario presented in Figure 6.3 the maximum target range is equal to:

$$R_{\max} = Htg\left(\gamma + \frac{\theta_v}{2}\right) \tag{6.20}$$

so the inequality (6.19) can be expressed as:

$$PRF < \frac{c}{2H} ctg\left(\gamma + \frac{\theta_v}{2}\right) \tag{6.21}$$

In the SAR configuration presented in Figure 6.3, there are no echoes for delays smaller than the delay corresponding to the minimal target distance $R_{\min}$:

$$R_{\min} = Htg\left(\gamma - \frac{\theta_v}{2}\right) \tag{6.22}$$

In such cases it is sometimes possible to select a higher PRF, exceeding the limit (6.21). If the range span of the imaged zone is smaller than the minimum target range ($R_{\max} - R_{\min} << R_{\min}$), then it is possible to send several pulses before the first one is received. When such a mode is used, the PRF is limited by the time delay corresponding to the difference between the maximum and minimum target distance:

$$PRF < \frac{c}{2\left(R_{\max} - R_{\min}\right)} = \frac{c}{2H\left(tg\left(\gamma + \frac{\theta_v}{2}\right) - tg\left(\gamma - \frac{\theta_v}{2}\right)\right)} \tag{6.23}$$

For narrow-beam antennas, inequality (6.23) can be simplified to:

$$PRF < \frac{c\cos^2(\gamma)}{2H\theta_v \sin(\gamma)} = \frac{c\lambda\cos^2(\gamma)}{2HW_a \sin(\gamma)} \tag{6.24}$$

Designers of SAR radars have to select a proper antenna size and a proper PRF, and must simultaneously satisfy the constraints (6.24) and (6.18):

$$\frac{2v}{L_a} < PRF < \frac{c\lambda\cos^2(\gamma)}{2HW_a \sin(\gamma)} \tag{6.25}$$

For low velocity, low-resolution (using a long antenna) SAR systems, the above constraint gives relatively wide zones of available PRF. However, high-speed satellite SAR systems can have a very narrow zone for PRF. The increase of SAR resolution is then limited by the constraint (6.25), and more sophisticated methods have to be used to overcome this problem. One possible method is the use of a multibeam (multisegment) antenna [23]. Another possibility is to use continuous noise waveforms, which are free of range and Doppler ambiguity.

For a satellite platform the orbit velocity is strictly connected with the orbit heights. Assuming a circular orbit the satellite velocity can be calculated from a well-known formula:

$$v = \sqrt{\frac{GM}{R_E + H}} \tag{6.26}$$

where G is the gravity constant, M is the mass of the Earth, and RE is the Earth's radius. For a typical low SAR orbit (H = 300–700 km), the orbiting velocity is close to 8 km/s. The relative speed of the platform is further modified by the rotation of the Earth and other phenomena, but in a simplified case it can be assumed to be constant.

The maximum pulse repetition frequency is upper-limited by the width of the strip observed by the radar [see constraint (6.24)]. For example, for the orbit height of 500 km, antenna beamwidth of 5°, and looking angle of 45°, the maximum PRF is limited to approximately 2 kHz. The range resolution of the image can be improved by using a high-bandwidth transmitted signal. To obtain a 10m ground resolution, the bandwidth of 15 to 25 MHz has to be used. Using modern microwave and digital technology, it is possible to generate and receive 100-MHz to 2-GHz bandwidth signals, so it is possible to obtain a tenths-of-centimeter range resolution in modern SAR systems. It is more difficult to obtain a high cross-range resolution. The first limitation comes from the power budget. To obtain the 1m resolution, a relatively small antenna (2m long) has to be used. This type of antenna has relatively low gain, and thus a high power of transmitted signal is required. The second limitation comes from the received signal Doppler bandwidth, predicted by (6.11). The pulse repetition frequency is then low-limited by constraint (6.18). For the 1m cross-range resolution (2m antenna length), the PRF must be higher than 8 kHz. For the 10-cm resolution, the PRF must be higher than 80 kHz. In both presented examples the two constraints (6.18) and (6.24) exclude themselves mutually, so it is not possible to obtain a high cross-range resolution using the classical space SAR strip mode. One of the possible solutions to obtain a high cross-range resolution is to decrease the vertical width of the antenna beam. As a consequence, it would create a very narrow scanning strip. Another solution is to use the spotlight mode [24–27] instead of the strip mode. In the spotlight mode the SAR antenna is directed to one point on the Earth's surface, changing its position in satellite coordinates. This mode of operation leads to the increase of the observation time, which consequently leads to the increase of the Doppler bandwidth of the received echo signal. This is equivalent to an increase of the cross-range resolution of the final image. Except for this significant advantage (increase of cross-range image resolution), spotlight technology has many disadvantages. It requires high-precision mechanical or electronic antenna beam steering. The imaged area is relatively small, and signal processing is much more complicated as it is usually necessary to take into account the higher-order terms in (6.5) and compensate the range walks of the observed points.

The ambiguity problems discussed above exist only in the classical pulse SAR. The random or the quasi random continuous waveforms are free from the ambiguity problem [28], and it is possible to design a high-resolution SAR using noise radar technology.

6.2 Noise SAR

One of the alternatives to the classical pulse SAR is the continuous-wave noise synthetic aperture radar. This radar simultaneously transmits and receives high-frequency, narrowband noise, or a noise-like signal [29–35]. Such a signal is usually synthesized using noise amplitude or a phase modulation of the carrier sine signal. The transmitted signal ambiguity function, in contradiction to the pulse signal, has only one main lobe, and thus no ambiguity problem is present in this type of radar [28]. The PRF limitations (6.25) are no longer valid during the design and manufacturing of SAR, and it is possible to develop a spaceborne, high-resolution SAR satellite with a wide scanning strip.

An example of the SAR scenario is presented in Figure 6.6. The platform is moving along the Y-axis at the height H and the speed v. Let us consider a single scatterer placed at the position $(X_o, Y_o, 0)$ on the surface of the Earth. The distance between the electromagnetic center of the SAR antenna and the point scatterer can be expressed by the formula:

$$r(t) = \sqrt{H^2 + X_o^2 + (vt - Y_0)^2} = \sqrt{R_o^2 + (vt - Y_0)^2} \tag{6.27}$$

Let us assume that the radar is sending a narrowband noise signal with the complex envelope $x_T(t)$ at the carrier frequency F. The received signal $x_R(t)$, originating from the analyzed point scatterer is an attenuated, time and Dop-

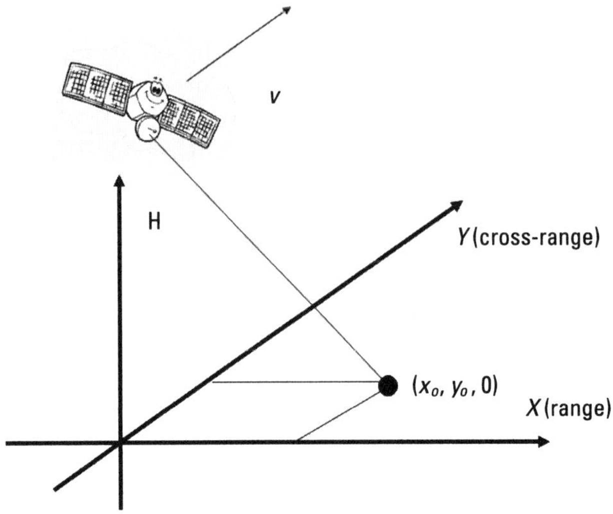

Figure 6.6 The scenario of satellite synthetic aperture radar imaging.

pler-shifted copy of the transmitted signal multiplied by the time-windowing function $w^2\left(\frac{vt - Y_o}{R_o}\right)$ and can be expressed by the formula:

$$x_R(t) = Ax_T\left(t - \frac{2r(t)}{c}\right)\exp\left(-j\frac{4\pi r(t)F}{c}\right)w^2\left(\frac{vt - Y_o}{R_o}\right) \tag{6.28}$$

where A is echo signal amplitude, depending on the range to the target, transmitted signal power, antenna gain, and total losses, and w(α) denotes the along-track antenna angular radiation pattern (gain related to the maximum gain).

The final SAR image creation in a noise SAR is more complicated than with a pulse SAR. Noise SAR processing is based not on the matched filtering concept but on correlation signal processing. The final SAR image is calculated in a correlation processor. This processor correlates the single-dimensional input data stream with the signal model (6.28). The correlation procedure is described by the formula:

$$I(x_o, y_o) = \int_{t=Y_0/v - t_i/2}^{Y_0/v + t_i/2} x_R(t) x_T^*\left(t - \frac{2\sqrt{H^2 + X_o^2 + (vt - Y_o)^2}}{c}\right) \exp\left(j\frac{4\pi\sqrt{H^2 + X_o^2 + (vt - Y_o)^2}\,F}{c}\right) w^2\left(\frac{vt - Y_o}{\sqrt{H^2 + x_o^2}}\right) dt \tag{6.29}$$

where $t_i = T_{ob} = \sqrt{\dfrac{H^2 + X_o^2}{2L_a v}}$ is the integration time, depending on the target range.

While the mean distance between the satellite and the target does not fulfill the constraint (6.13), the space SAR image formation has to take into account the target range migration.

The computational complexity of (6.29) is much higher than the computational complexity of classical pulse SAR processing, but in many cases the space SAR image formation is performed off-line in ground stations, so the computational power is not the main limitation of the spaceborne SAR system. In many practical cases (6.29) can be simplified to decrease the computational complexity of the image formation.

The verification of the presented method was done using simulations. Figure 6.7 presents an example of a noise SAR image of a scene with five reflecting points forming a cross. The range and cross-range resolution is equal to the resolution of a classical SAR.

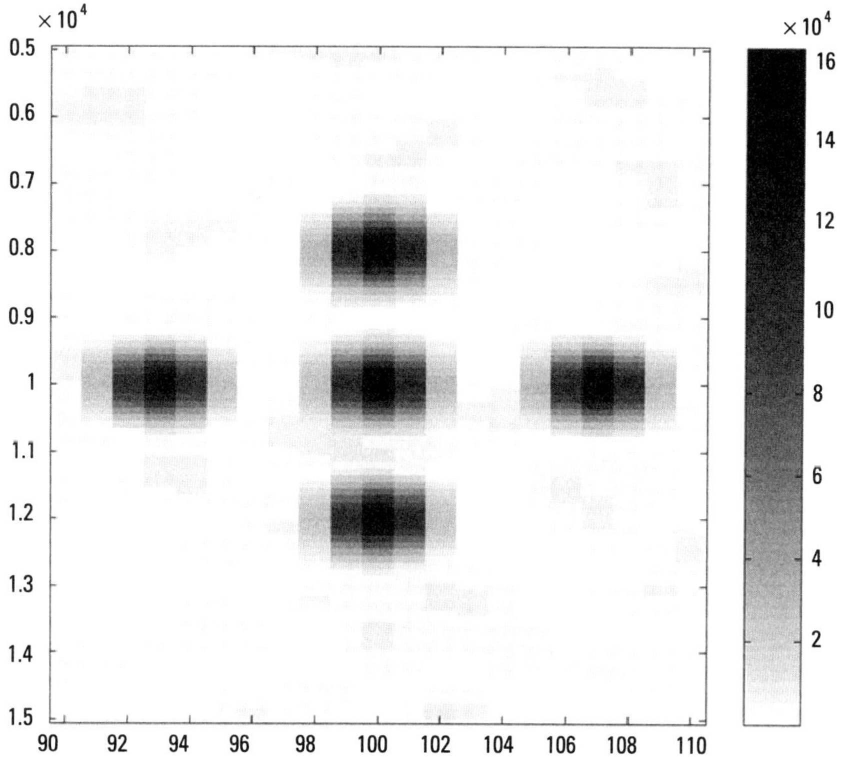

Figure 6.7 The final SAR image; *x*-axis – range cells, *y*-axis – samples (time). Grayscale: amplitude of SAR image.

The raw radar data was created using a Raw Radar Data Simulator [36, 37]. The magnitude of the radar return is presented in Figure 6.8. It is easy to notice that the return signal has the antenna beam modulation and the additional modulation caused by the interferences between five echo components. The raw radar data, in contradiction to the pulse SAR mode, does not have any range resolution and the cross-range resolution is equal to the antenna beamwidth at the selected distance. It is worth noting that in this case the signals which originated from all the scattering points from different ranges overlap. All useful information is hidden in the signal complex spectrum. The magnitude of this spectrum is presented in Figure 6.9. Due to the interferences between several echo components the received signal spectrum differs from the transmitted signal spectrum presented for reference in Figure 6.10.

6.3 Adaptive SAR Image Enhancement

The image formation procedure described by (6.29) is not optimal in the case where strong reflecting points and weak scatterers are simultaneously present on

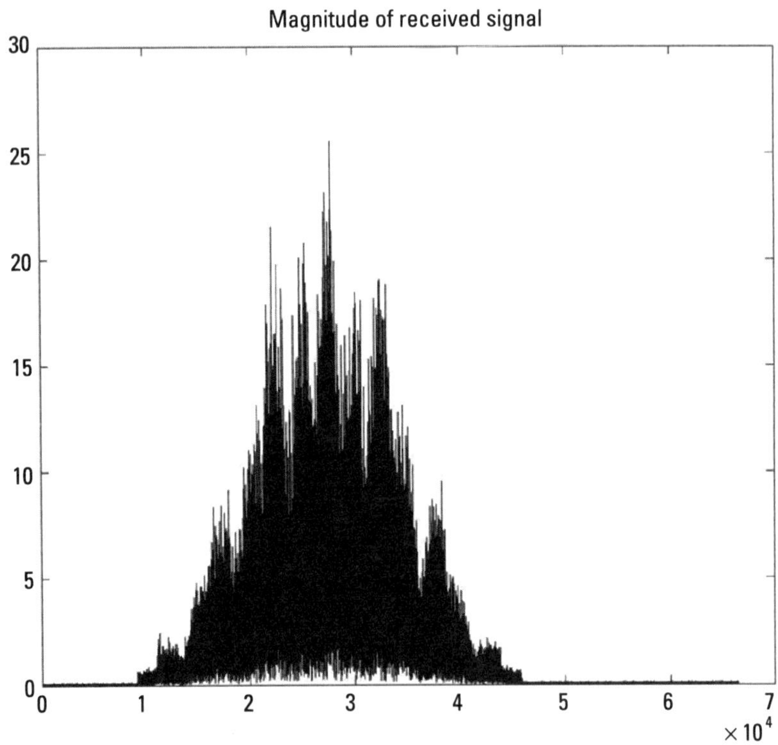

Figure 6.8 Magnitude of the return signal, *x*-axis – samples.

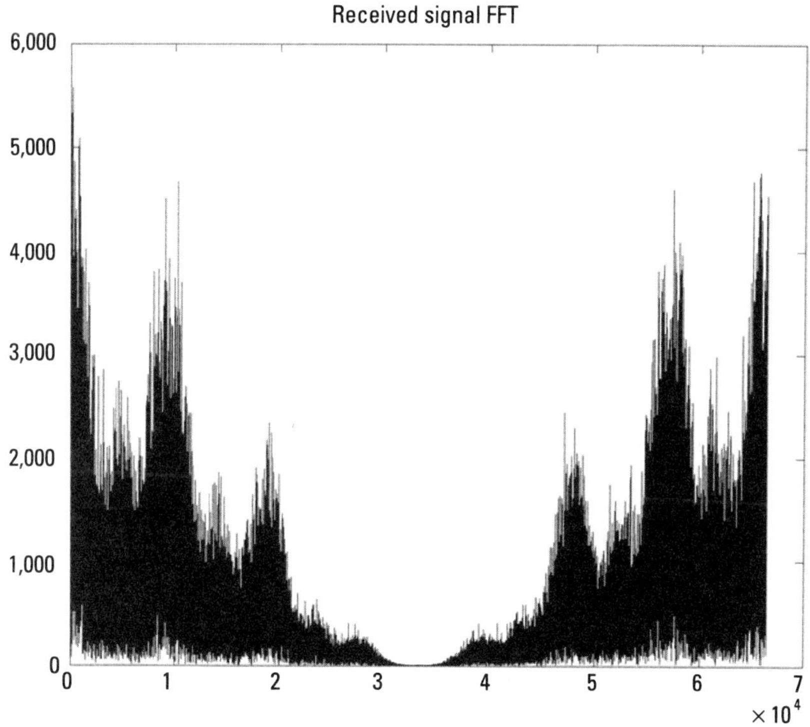

Figure 6.9 Magnitude of the spectrum of the return signal, *x*-axis – FFT bins.

the scene. In this case the echo originating from a strong scatter can mask the weak targets [38], and the overall picture quality is poor. The mean level of the echo sidelobes is smaller than the main lobe by the factor Bt_i, where B is signal bandwidth and t_i is the integration time. Although the Bt_i factor in space SAR can have a value in the range of 60to 80 dB, the masking problem can appear when the dominant scatterer is 30 to 50 dB above the mean scatterer level, due to the fact that all scatterers contribute to the total noise floor of the SAR image. One possible solution of this problem is the use of the adaptive image-enhancing algorithm based on the adaptive cancellation of the main scatterers' echoes [38–41] (CLEAN method).

Let us consider that the received signal consists of several strong echoes and many weaker ones. In the first step of the algorithm the primary SAR image is formatted using a correlation algorithm (6.29). In the second stage the main scatterers are detected, and for each one its position $(\hat{X}_k, \hat{Y}_k)$ and complex amplitude $\hat{A}_k$ are estimated. On the basis of the signal model (6.28), the scatterer raw signal $\hat{x}_k(t)$ is modeled. This signal is in the form of:

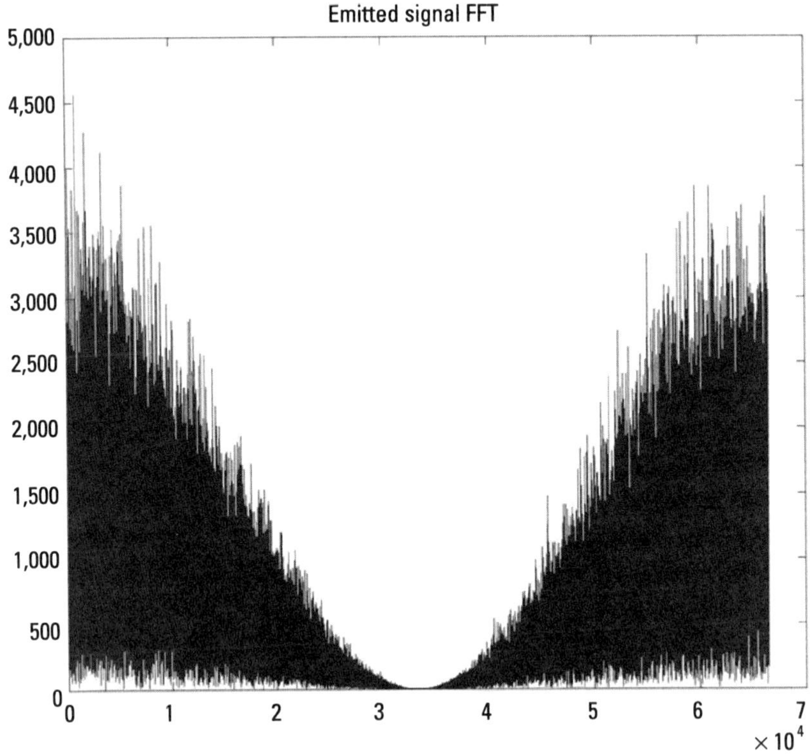

Figure 6.10 Magnitude of the transmitted signal spectrum, *x*-axis – FFT bins.

$$\hat{x}_k(t) = \hat{A}_k x_T^* \left(t - \frac{2\sqrt{H^2 + \hat{X}_k^2 + (vt - \hat{Y}_k)^2}}{c} \right) \exp\left(j\frac{4\pi\sqrt{H^2 + \hat{X}_k^2 + (vt - \hat{Y}_k)^2}\, F}{c} \right) w\left(\frac{vt - \hat{Y}_k}{\sqrt{H^2 + \hat{X}_k^2}} \right) \tag{6.30}$$

In the third step of the algorithm, the modeled signal is subtracted from the received signal:

$$x_R^{(k)}(t) = x_R^{(k-1)}(t) - \hat{x}_k(t) \tag{6.31}$$

where $x_R^{(0)}(t) = x_R(t)$ is the received signal.

In the fourth algorithm step the weak target SAR image with removed strong scatterers is created. This image is formed using signal (6.31) for the correlation procedure. Steps 2 to 4 can be repeated until no strong dominant scatterers are detected in the SAR image.

In the last stage of the algorithm, the final image is formatted by selecting the strongest points detected from the primary image (from the first algorithm stage or from the fourth stage if the repetition of the strongest target removal is applied) and incorporating them into the image created in the previous stage of the algorithm.

The results of the simulations carried out by the author are presented in Figures 6.11 through 6.14. In the first step of the experiment the Raw Radar Data Simulator [36, 37] was used to generate a reflected signal from five reflecting points, using noise illumination. Each point has the same radar cross section (100 m^2) and the points form a cross on the *X*-*Y* plane. The SAR image obtained using correlation processing (6.29) is presented in Figure 6.11. The processing noise floor is 40 dB below the strongest echo. The time-bandwidth product is equal to 40,000.

This high level of noise floor can mask weak echoes. In the next step of the simulation the cross-section of the central scatterer was decreased by 40 dB (to 0.01 m^2). Figure 6.11 presents the SAR image obtained by applying correlation

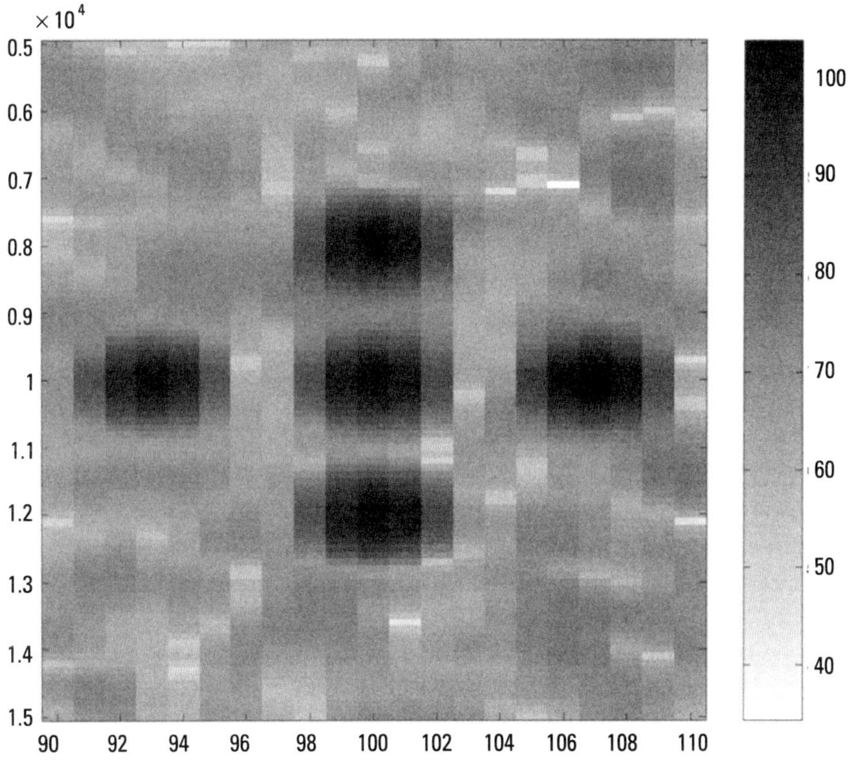

Figure 6.11 Example of a noise SAR image, five identical scatterers, *x*-axis: range, *y*-axis: cross-range, grayscale in decibels (as in Figure 6.10).

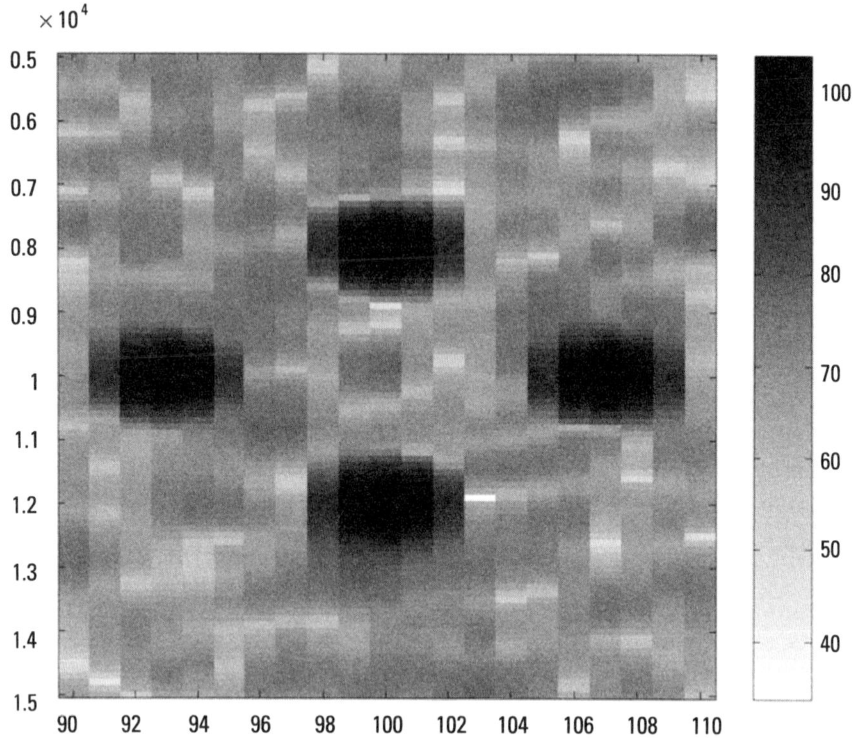

Figure 6.12 Example of a noise SAR image, four identical strong scatterers (20 dBsm) and one weak (central −20 dBsm) scatterer. The central scatterer (40 dB smaller than side scatterers) is hidden by the processing noise floor (sidelobes). *x*-axis: range, *y*-axis: cross-range, grayscale in decibels.

processing. The high sidelobes of the strong side scatterers mask the central scatterer echo, which is not visible in this case (as it is hidden in the strong scatterers' sidelobes).

The results of applying the adaptive SAR image enhancement algorithm are presented in Figures 6.13 and 6.14. The SAR image after the dominant echoes removal procedure is presented in Figure 6.13. The central scatterer is now clearly visible and the residuals of the dominant scatterer echoes are also present. The removal procedure attenuates the strongest echoes by 50 to 60 dB. The final SAR image is presented in Figure 6.14. All scatterers are visible and the processing noise floor has been decreased by 40 dB. The procedure described can be further applied to the weakest echoes, and a further decrease in the noise floor by a few decibels can be achieved.

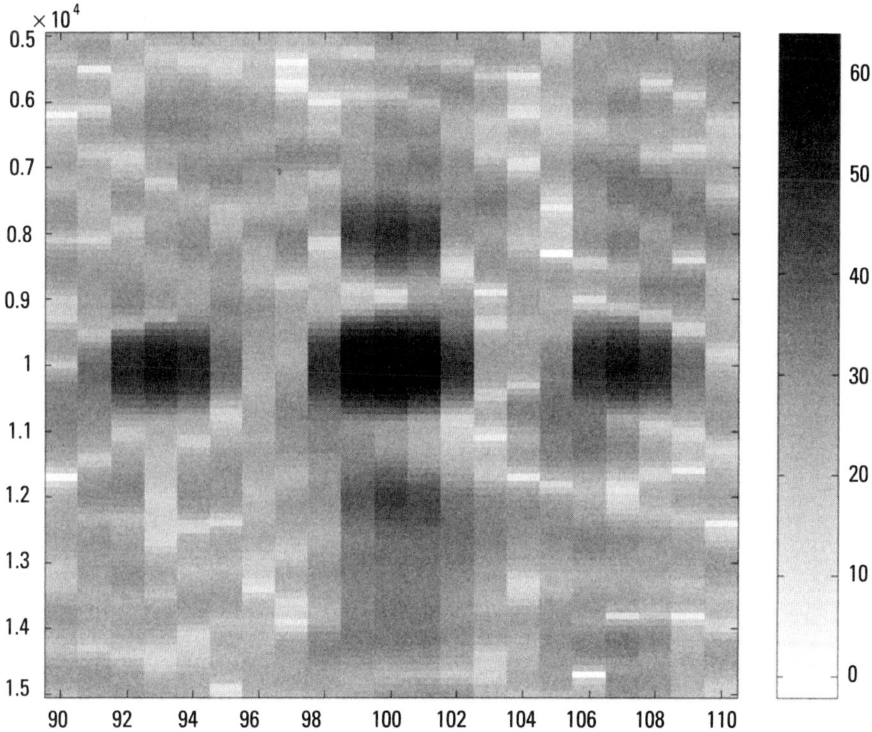

Figure 6.13 SAR image after the dominant echoes removal procedure. The central scatterer is clearly visible, and the side scatterer echoes significantly attenuated (50–60 dB). *x*-axis: range, *y*-axis: cross-range, grayscale in decibels.

6.4 Ground-Based SAR Example

In this chapter an example noise SAR demonstrator built in the Radar Technology Laboratory of the Warsaw University of Technology is presented. The aim of the demonstrator's construction was to gain experience with the noise and pseudo-noise signals applied in SAR technology.

The SAR demonstrator was built using commercial-off-the-shelf (COTS) elements: an arbitrary waveform generator (AWG, Agilent MXG N5182A), a directive coupler, two antennas (Tx and Rx), and a two-channel vector signal analyzer (VSA, Agilent 89600). The main components of the noise SAR demonstrator are presented in Figure 6.15.

The arbitrary waveform generator (AWG) can modulate the carrier frequency up to 3 GHz with an arbitrary complex baseband waveform uploaded to the internal memory of the instrument. For the purposes of the test, Gaussian white noise was generated using the MATLAB software and downloaded to the AWG. The AWG is equipped with two digital-analog converters (for

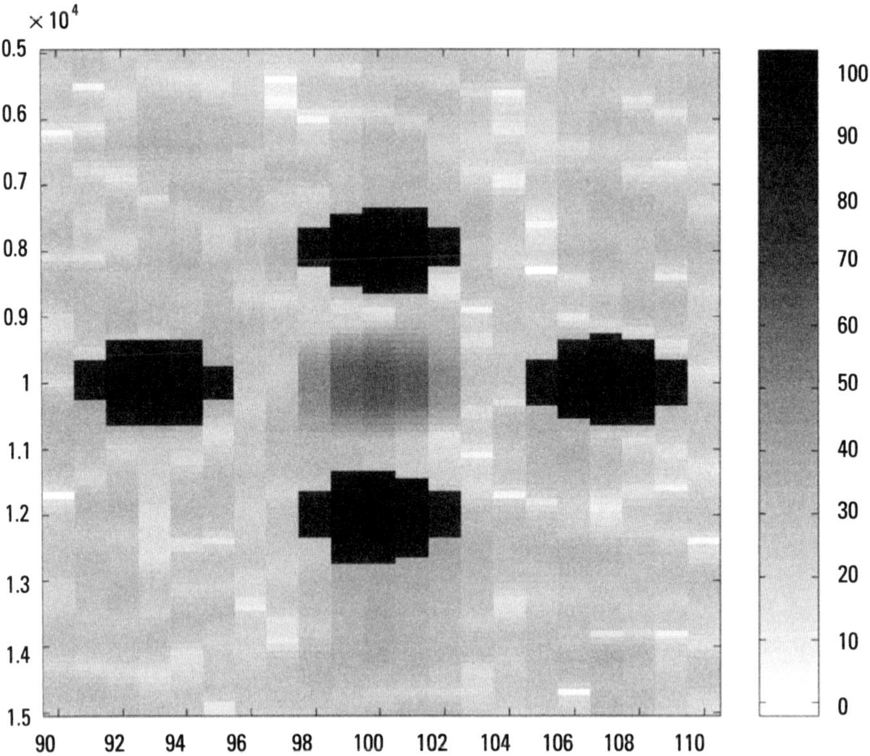

Figure 6.14 Final SAR image after CLEANing and dominant echo reinsertion, *x*-axis: range, *y*-axis: cross-range, grayscale in decibels.

I and Q channels) working at a clock frequency of 60 MHz; therefore, the theoretical signal bandwidth is almost 60 MHz. The sampling frequency of the Vector Signal Analyzer (VSA) is about 46 MHz (again I and Q sampling) and the maximum signal bandwidth is equal to 36 MHz. This limitation comes from antialiasing filters used in the VSA. The VSA is equipped with two channels. The first VSA channel was used to record the surveillance signal from the receive antenna and the second was used to record the reference (transmit) signal, as shown in Figure 6.16. The signal generated by the AWG generator was split into two parts. The main part of the energy was directed to the transmit antenna, and a small portion (–20 dB) was used as the reference signal. Both channels of the VSA have a common local oscillator (LO) and synchronized sampling clock so both signals were recorded coherently. The only custom element of the demonstrator hardware was a movable antenna platform, used to obtain the movement of the SAR sensor.

The experiment was carried out on a balcony of the Radar Technology Radiolocation and Digital Signal Processing Students' Research Laboratory, located about 21m above the ground surface. The imaged area was the space

Figure 6.15 Noise SAR demonstrator hardware.

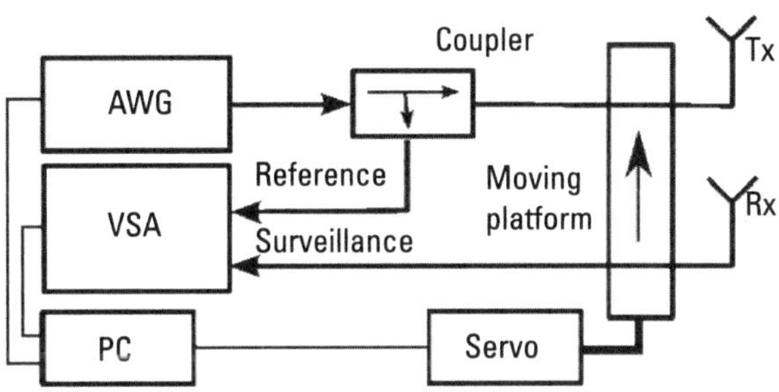

Figure 6.16 Block diagram of the noise SAR demonstrator.

between the building of the Faculty of Electronics and Information Technology, on which the system was placed, and the building of the Faculty of Civil Engineering, situated parallel and at a distance of about 110m.

The SAR sensor consisted of two widebeam patch antennas, mounted to a platform moving along the balcony's railing. Accurate positioning of the platform was possible due to a computer-controlled step motor. The system worked in a stop-and-go mode, which means that each sounding was obtained from a stationary sensor position and there was no signal distortion caused by the Doppler effect. The carrier frequency was set to 1.920 GHz in a single-band mode while the center frequency of the stepped-frequency recordings was equal to 1.992 GHz. To satisfy the Nyquist theorem and avoid aliasing in the azimuth domain, the step between the sensor's positions had to be smaller than $\lambda/4$, where λ is the carrier wavelength. As a result the scene was sounded with a 3-cm step between sensor positions, which gave 300 soundings positions and a 9-m synthetic aperture length. A single recording size was 2^{19} samples, which was equivalent to a 10-ms emission. The Rx and Tx antennas were separated by 1.3m, and the crosstalk signal between the antennas was about 30 dB higher than the strongest scene echoes. The crosstalk between the transmit and receive antennas is unwanted from the point of view of the SAR image formation, since it raises the correlation sidelobes. For this reason the crosstalk was removed from the surveillance channel using the lattice filter described in detail in Chapter 4.

The radar demonstrator was used in two modes: single band and multiband. In the single band the range resolution was about 5m. As the demonstrator used a stop-and-go antenna motion, it was possible to divide all signal processing into two stages: range compression and cross-range compression.

The first stage of the stop-and-go SAR processing was range compression. Due to the random character of the sounding signal, range compression was performed for the noise radar by cross-correlating the received signal (after CLEANing) with the reference signal. An efficient way to compute the cross-correlation between the reference and surveillance signals is to make a computation in the frequency domain described. As a results of the calculation of the correlation function, complex range profiles were obtained. The next step is the final SAR image formation using a selected method of cross-range compression, which in this case was the simplest backpropagation method [8] due to limited aperture length. The image is formed by the calculation of a complex pixel value for each node of a rectangular grid in the x, y coordinates, where x corresponds to azimuth and y corresponds to range. For all nodes of the grid the following processing is applied: first, for each antenna position (indexed with n), the bistatic distance $r_{x,y}(n)$ from the Tx antenna to the grid node (x, y) and back from the node to the Rx antenna is computed as:

$$r_{x,y}(n) = \sqrt{(x_{Tx}(n) - x)^2 + y^2} + \sqrt{(x_{Rx}(n) - x)^2 + y^2} \tag{6.32}$$

where y is the fast range coordinate of the node and x_{Tx}, x_{Rx}, and x are, respectively, the slow range coordinates of the nth Tx antenna position and nth Rx antenna position and the node. Then the time delay of the expected echo from the node can be calculated for each antenna position as:

$$d_{x,y}(n) = d_0 + \frac{r_{x,y}(n)}{cT_s} \tag{6.33}$$

where d_0 is the constant delay resulting from the system cables length and T_s is the effective sampling period.

With this information each range profile can be shifted so that the expected response of the imaged point is placed exactly in the first column of the profile matrix. This procedure is called range cell migration correction (RCMC) and can be performed by the modulation of each profile's spectrum with $\exp(-j\theta d_{x,y}(n))$, which is equivalent to a shift in the time domain by $d_{x,y}(n)$ samples. Having the expected point's response collected in one vector, matched filtering in azimuth is performed. Its aim is to detect whether the energy gathered in the vector comes from the examined scene point. The filtering is based on phase differences of the point's response caused by different distances between the point and subsequent antenna positions. Expected phases of the point response may be expressed as:

$$\phi_{x,y}(n) = \frac{2\pi}{\lambda} r_{x,y}(n) \tag{6.34}$$

and the image pixel value can be calculated as:

$$A_{x,y} = \sum_{n=1}^{N} \exp(-j\phi_{x,y}(n)) R_{x,y}(n) W(n) \tag{6.35}$$

where $A_{x,y}$ is the value of the complex pixel intensity of the SAR image, N is the total number of antenna positions, $R_{x,y}(n)$ is a correlation sample from the nth antenna position after RCMC, and $W(n)$ is a Hamming window function. The absolute value of A is the SAR image's pixel brightness.

To increase the radar range resolution, the step stepped frequency approach was used. In the presented case five subbands, each with a bandwidth of 36 MHz, were used, giving a total bandwidth of 180 MHz. This corresponds to a 0.8m range resolution, similar to the expected cross-range resolution.

There are two methods of band synthesis: summation of interpolated signals in the time domain and the spectrum concatenation. The second approach was used as it is computationally more effective. The process of the band synthesis is very simple: from each band a useful part of the spectrum is extracted (cut) and placed in a proper position in the final spectrum. After obtaining synthetized spectra of the transmit and receive signals, the cross-correlation function is calculated by computing the inverse FFT from the product of the spectra.

The resulting SAR image obtained in the multiband mode is presented in Figure 6.17. For reference the optical image of the scene is presented in Figure 6.18. It is clearly seen that all the major elements of the real scene are visible on

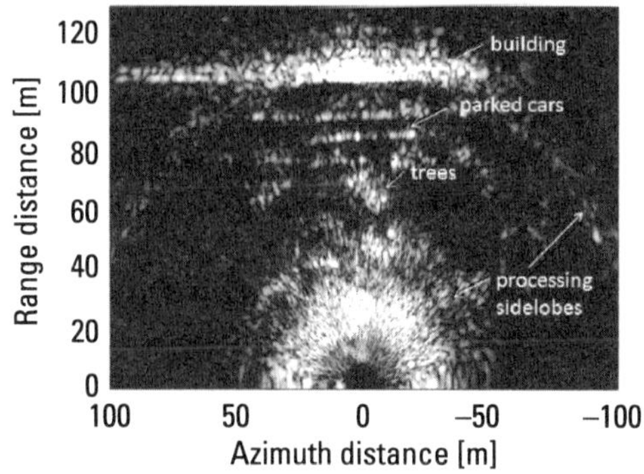

Figure 6.17 SAR image in multiband mode.

Figure 6.18 Optical image of the scene.

the SAR image. In addition to the large building, rows of parked cars, groups of trees, and the smaller building are also clearly visible. Even a rectangular metal fence around the small building can be observed in the image. The ring in the near distance is produced by the noise sidelobes.

References

[1] Kim, K. H., S. G. Kim, and J. W. Yi, "Detection of Ship Targets Near Coastline by Using Doppler Beam Sharpening Technique," *2011 3rd International Asia-Pacific Conference on Synthetic Aperture Radar (APSAR)*, September 26–30, 2011, pp. 1–4.

[2] Picardi, G., R. Seu, and S. Sorge, "Extensive Non-Coherent Averaging in Doppler Beam Sharpened Space-Borne Radar Altimeters," *1998 IEEE International Geoscience and Remote Sensing Symposium Proceedings (IGARSS '98)*, Vol. 5, July 6–10, 1998, pp. 2643–2645.

[3] Cassabaum, M. L., et al., "Feature Analysis Using Millimeter-Wave Real Beam and Doppler Beam Sharpening Techniques," *Proceedings Fifth IEEE Southwest Symposium on Image Analysis and Interpretation 2002*, 2002, pp. 101–105.

[4] Pietrzyk, G., et al., "Real-Time Implementation of Doppler Beam Sharpening Technique with Simple Motion Estimation," *First European Radar Conference 2004 (EuRad)*, October 11–15, 2004, pp. 185–188.

[5] Skolnik, M. I., *Radar Handbook*, 3rd ed., New York: McGraw-Hill, 2008.

[6] Franceschetti, G., and R. Lanari, *Synthetic Aperture Radar Processing*, Boca Raton, FL: CRC Press, 1999.

[7] Oliver, C., and S. Quegan, *Understanding Synthetic Aperture Radar Images*, Raleigh, NC: SciTech Publishing, 2004.

[8] Carrara, W. G., R. M. Majewski, and R. S. Goodman, *Spotlight Synthetic Aperture Radar: Signal Processing Algorithms*, Norwood, MA: Artech House, 1995.

[9] Sherwin, C. W., P. Ruina, and R. D. Rawcliffe, "Some Early Developments in Synthetic Aperture Radar Systems," *IRE Trans.*, Vol. MIL-6, April 1962, pp. 111–115.

[10] Cutrona, L. J., et al., "A High Resolution Radar Combat-Surveillance System," *IRE Trans.*, Vol. MIL-5, April 1961, pp. 127–131.

[11] Greene, C. A., and R. T. Moller, "The Effect of Normally Distributed, Random Phase Errors on Synthetic Array Gain Patterns," *IRE Trans.*, Vol. MIL-6, April 1962, pp. 130–139.

[12] Wiley, C. A., "Pulsed Doppler Radar Methods and Apparatus," U.S. Patent, No. 3,196,436, Filed August 1954, 1965.

[13] Wehner, D. R., *High Resolution Radar*, 2nd ed., Norwood, MA: Artech House, 1995.

[14] Curlander, J. C., and R. N. McDonough, *Synthetic Aperture Radar Systems and Signal Processing*, New York: John Wiley & Sons, 1991.

[15] Dawidowicz, B., et al., "First Polish SAR Trials," *IEE Proc.-Radar, Sonar & Navig.*, Vol. 153, No. 2, April 13, 2006, pp. 135–143.

[16] Purchla, M., and M. Malanowski, "Simple Motion Compensation Algorithm for Unfocused Synthetic Aperture Radar," *XII Wilga Symposium IEEE-SPIE*, Wilga, Poland, 2003.

[17] Kulpa, K., M. Purchla, and M. Malanowski, "Real-Time Unfocused SAR Algorithm with Motion Compensation," *International Radar Symposium (IRS-2003)*, Dresden, Germany, 2003.

[18] Kulpa, K., et al., "The Use of Autofocus SAR Algorithm in Platform Velocity Estimation," *International Radar Symposium (IRS-2004)*, Warsaw, Poland, 2004.

[19] Samczynski, P., and K. S. Kulpa, "Coherent MapDrift Technique," *IEEE Transactions on Geoscience and Remote Sensing*, Vol. 48, No. 3, March 2010, pp. 1505–1517.

[20] Samczynski, P., "Superconvergent Velocity Estimator for an Autofocus Coherent MapDrift Technique," *IEEE Geoscience and Remote Sensing Letters*, Vol. 9, No. 2, March 2012, pp. 204–208.

[21] Samczynski, P., "The Coherent MapDrift Technique Robustness to Low SNR SAR Data," *2008 Microwaves, Radar and Remote Sensing Symposium (MRRS 2008)*, September 22–24, 2008, pp. 240–243.

[22] F. Berizzi, E. Dalle Mese, and M. Martorella, "Performance Analysis of a Contrast-Based ISAR Autofocusing Algorithm," *Proceedings of the 2002 IEEE Radar Conference*, Long Beach, CA., April 2002, pp. 200–205.

[23] Koch, P., et al., "Dual-Polarized Feed-Cluster for a Reflector-Based Multi-Beam SAR Antenna," *Proceeding of 17th International Crimean Conference Microwave & Telecommunication Technology (CriMiCo 2007)*, 2007.

[24] Liu, F., et al., "Performance Analysis of Two-Step Algorithm in Sliding Spotlight Space-Borne SAR," *2010 IEEE Radar Conference*, May 10–14, 2010, pp. 965–968.

[25] Preiss, M., D. Gray, and N. Stacy, "Space Variant Filtering of Polar Format Spotlight SAR Images for Wavefront Curvature Correction and Interferometric Processing," *2002 IEEE International Geoscience and Remote Sensing Symposium 2002 (IGARSS '02)*, Vol. 1, 2002, pp. 179–181.

[26] Breit, H., et al., "TerraSAR-X Payload Data Processing—First Experiences," *IEEE International Geoscience and Remote Sensing Symposium 2007 (IGARSS 2007)*, July 23–28, 2007, p. 3936.

[27] Torre, A., D. Calabrese, and M. Porfilio, "COSMO-SkyMed: Image Quality Achievements," *2011 5th International Conference on Recent Advances in Space Technologies (RAST)*, June 9–11, 2011, pp. 861–864.

[28] Axelsson, S. R. J., "Suppressed Ambiguity in Range by Phase-Coded Waveforms," *IEEE 2001 International Geoscience and Remote Sensing Symposium (IGARSS'01)*, Vol. 5, July 9–13, 2001, pp. 2006–2009.

[29] Axelsson, S. R. J., "On the Theory of Noise Doppler Radar," *Proc. IGARSS 2000*, Honolulu, HI, July 24–28, 2000, pp. 856–860.

[30] Garmatyuk, D., and R. M. Narayanan, "Ultrawide-Band Noise Synthetic Radar: Theory and Experiment," *IEEE Antennas Propagat. Soc. Int. Symp. 1999*, Vol. 3, Orlando, FL, July 1999, pp. 1764–1767.

[31] Lukin, K., P. Vyplavin, and N. Kolchigin, "Noise Waveform SAR Imaging in Antenna Near Zone," *International Radar Symposium 2006*, Krakow, Poland, May 24–26, 2006, pp. 303–306.

[32] Lukin, K., et al., "2D and 3D Imaging Using S-Band Noise Waveform SAR," *2011 3rd International Asia-Pacific Conference on Synthetic Aperture Radar (APSAR)*, September 26–30, 2011, pp. 1–4.

[33] Kulpa, K., et al., "Quality Enhancement of Image Generated with Bistatic Ground Based Noise Waveform SAR," *IET Radar, Sonar & Navigation*, Vol. 2, No. 4, August 2008, pp. 263–273.

[34] Maslikowski, L., and J. S. Kulpa, "Noise SAR Using Waveform with Reduced Correlation Noise Floor," *2011 European Radar Conference (EuRAD)*, October 12–14, 2011, pp. 218–221.

[35] Maslikowski, L., M. Malanowski, and K. Kulpa, "Stepped Frequency Ground-Based Noise SAR Demonstrator," *2011 3rd International Asia-Pacific Conference on Synthetic Aperture Radar (APSAR)*, September 26–30, 2011, pp. 1–2.

[36] Kulpa, K., A. Gorzelanczyk, and M. Smolarczyk, "Radar Signal Generator and Its Usage for SAR Algorithm Tests," *MIKON 2002 XIV International Conference on Microwaves, Radar and Wireless Communications*, Gdansk, Poland, May 20–22, 2002.

[37] Gromek, D., et al., "SAR/InSAR Raw Data Simulator Using DTM Scene Definitions," *2012 13th International Radar Symposium (IRS)*, 2012, pp. 153–156.

[38] Kulpa, K. S., and Z. Czekala, "Masking Effect and Its Removal in PCL Radar," *IEE Proceedings Radar, Sonar and Navigation*, Vol. 152, No. 3, June 3, 2005, pp. 174–178.

[39] Kulpa, K., and Z. Czekala, "Short Distance Clutter Masking Effects in Noise Radars," *J. Applied Radio Electronics*, Vol. 4, No. 1, 2005, pp. 96–98.

[40] Kulpa, K. S., and Z. Czekała, "Ground Clutter Suppression in Noise Radar," *Proc. Int. Conf. RADAR 2004*, Toulouse, France, October 18–22, 2004, p. 236.

[41] Kulpa, K., et al., "Quality Enhancement of Image Generated with Bistatic Ground Based Noise Waveform SAR," *IET Radar, Sonar & Navigation*, Vol. 2, No. 4, August 2008, pp. 263–273.

7

Passive Detection of Moving Targets

In the previous chapters active radar technology was presented. Active pulse radars have been used for many years to detect moving targets [1]. Active radar technology is now established, but its high peak power makes active radars very easy to be detected and classify, and thus their expected lifetime on the modern battlefield is very limited. For a long time scientists and engineers have been searching for a radar technology that would have a low probability of intercept (LPI). The search for an LPI radar with a low peak power and a long detection range has led to the development of continuous-wave radars such as frequency modulated continuous-wave (FMCW) radars [1–8], ultrawideband radars, and noise radars [9–13]. This approach eliminates some of the disadvantages of active radars such as very high peak transmitted power. However, more covert sensors are still required due to the fast progress in electronic support measurements [7] and electronic countermeasures [14–20]. Most of the active radars can today be detected and jammed so that alternative technologies free of their own emissions are being investigated.

The sharp increase in available computational power makes it possible to develop the family of passive radars. The leading technology in this field is passive coherent location (PCL) [21–29], also referred to as passive covert radars (PCR) [30–34]. In this technology target detection is based on the determination of the range-Doppler correlation function between the signal used for the illumination of the target and its return echo. To illuminate the target, various types of emitters of opportunity can be used, depending on the availability, required detection range, and properties. The most popular systems exploit FM radio [27, 29, 35–40]. There have been many studies on the exploitation of analog television illumination [41–46], but this signal is obsolete nowadays. Many countries have already switched off this transmission and many more intend to do it over the next couple of years. Much more attention has been attracted by digital broadcasting. Digital video broadcasting from satellites (DVB-S) provides very wide coverage even at low altitude, and also provides very wide frequency bands (30 MHz in a single channel, more than 2 GHz in total [47–50]), but the power density is very low and thus practical use is very limited. A similar problem occurs in exploiting GPS and GLONASS signals for passive location [51–55]. Illumination in this case is worldwide, but the power density is also very low. To improve the target echo to noise signal, the forward scatterings concept has also been investigated [56, 57]. Terrestrial digital signals are much more attractive than those in satellites. Much research, along with practical application, has been done on terrestrial digital television (DVB-T) [58–66]. The DVB-T transmitters have significant power (up to 100 kW), a bandwidth comparable with active radars (8 MHz) and good coverage in order to provide a high-quality service to subscribers. This service is very well developed in many countries and is a good candidate for application in passive radars. Digital audio broadcasting transmitters (DAB) are also of interest as illuminators of opportunity [67–70], but the DAB signal is not as popular as it once was. The other group of signals exploited in passé radars are communication signals. Medium range passive radars (up to 50 km) can exploit cellular phone signals (GSM and UMTS among others) [71–80] and Worldwide Interoperability for Microwave Access (WiMAX) networks [81–85]. Low-power WiFi networks can also be used for passive sensing in short-range applications (up to several kilometers) [86–89].

The PCL technique can be named passive only from the sensors' point of view, but it still requires the existence of powerful illuminators of opportunity. Radar designers aspire to develop a fully passive system based on the radiation generated by the target. A wide scope of existing passive sensors called passive emitter tracking (PET) exploit the target's man-made radio emissions such as on-board radio and radar transmitters, data link transmitters, secondary radar returns, and others. Two technologies are commonly used; namely, the monostatic direction of arrival (DOA) method [90–92] and the multistatic time difference of arrival (TDOA) technique [93–93]. The multistatic method offers better accuracy than the monostatic one [98], but it requires much more computational power and broadband transmission links between sensor sites. However, this method works only when the onboard emission is switched on. A silent target can pass the defense line without being detected.

To detect targets in a situation when there are no man-made emissions, either on-board or off-board of the target, other physical phenomena must be taken into consideration. These phenomena should be inherent to the moving target of interest. One feature that is widely used for target detection is thermal emission of the body of the target. The hot surface emits a broadband noise signal in the high-frequency (HF), microwave, and infrared regions. Far infrared (thermal) cameras have been in widespread use for target detection and tracking in military and commercial systems for a long time [99–102]. Lower microwave frequencies are utilized in radiometric systems [103, 104] that find many fields of applications, such as radio-astronomy [105], remote sensing [106, 107], medicine [108], pollution control (detection of oil slicks on the sea surface) [109], mine detection [110], and emission detection [111]. In most applications, only the emitted signal power is used. In some cases, however, the phase information is also taken into consideration. Most detection algorithms in phase-sensitive radiometry are based on the examination of the cross-correlation function of signals received by the antenna sets [105]. More sophisticated algorithms are used also for high-resolution passive radar imaging based on SAR technology [107, 112].

In classical radiometry, it is not possible to detect targets with the same temperature and radiation properties as their surrounding environment. However, the usage of correlation radiometry enables detection of man-made moving targets. Until now, there have not been such commercially available systems, but the experiments on tracking oceanic streams or detecting atmospheric phenomena such as hydrometeors [107] show that it is possible to detect moving targets and estimate their position and velocity.

7.1 Radiometric Detection System

Thermal emission is a feature of all objects whose temperature is higher than 0 [K]. All man-made objects emit electromagnetic energy that can be used to detect and track such objects. Thermal emission covers a very wide frequency range, starting from the radio frequency and ending at the infrared or visible light frequency. According to Planck's law, the radiated power density of a perfectly black surface is described by the formula:

$$P = \frac{8\pi hF^3}{c^2 \exp\left(\frac{hF}{kT_t} - 1\right)} S_0 \tag{7.1}$$

where S_0 is the effective area of the target, k is the Boltzmann constant, T_t is the target temperature, c is the velocity of light, F is the frequency, and h is Planck's constant. The power, received by the narrowband microwave radiometric system with bandwidth B and central frequency F ($F >> B$), equipped with an antenna system with effective surface (effective aperture) Sr and gain Gr, at the distance r_t from the target, calculated using the Rayleigh-Jeans low-frequency approximation of Planck's law is equal to:

$$P_{tr} = \frac{S_o k T_t G_r}{2\pi r_t^2 L} B \tag{7.2}$$

where L stands for the receiver system losses. Several radiometry algorithms can be used for the detection of the target [111, 113, 114]. The most popular algorithm, very close to the optimal one, is based on the square-law detector and linear integration of the received power. The target detection range can be predicted as [115]:

$$r_{t_{\max}} = \sqrt{\frac{S_o T_t G_r \sqrt{t_i B}}{2\pi T_R D_0 L}}. \tag{7.3}$$

where D_0 is the detection threshold for a chosen probability of a false alarm and TR is the equivalent receiver temperature. The above equation determines the detection range in a clutter-free environment when only one target is present and detection is limited only by the receiver's system thermal noise. In most practical scenarios the target is detected against the ground clutter and/or the space noise, so the detection range is much shorter. The direct use of the radiometric concept does not allow a sensor to distinguish a moving target from the ground clutter, and moving targets can be undetected when either the classical radiometric system or infrared cameras are applied.

7.2 Detection of Moving Targets

The discrimination between the ground clutter and the moving target can be achieved using the Doppler frequency analysis. An example scenario of the passive detection of a moving target is presented in Figure 7.1.

The range velocity of the target (the change in time of the range between the target and the sensor) introduces the frequency shift to the narrow-band received signal. The Doppler frequency shift is equal to:

$$f_d = V_r / \lambda = V_r F / c \tag{7.4}$$

where $V_r = \frac{dr}{dt}$ is the target range velocity, λ is the wavelength, and F is the central frequency of the received signal. In active radars and PCL systems it is easy to detect even small Doppler frequency changes due to the fact that the reference (emitted) signal is available. In the case of radiometric systems it is not possible to calculate the Doppler shift from a single received signal. The

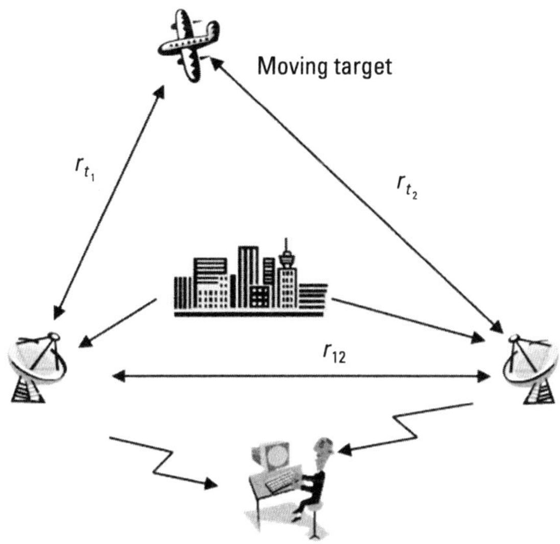

Figure 7.1 Example scenario of the passive detection of moving targets.

one and only way to estimate the Doppler frequency shift is to collect the signal radiated by the target using multiple receivers separated in space [115] and to perform correlation analyses of both received signals. The signal received by each receiver is a sum of the receiver's system thermal noise, ground clutter, and the target signal shifted in time and in frequency by the Doppler shift (7.4).

To perform a correlation process, both receivers must have identical phase and amplitude characteristics. This constraint can easily by fulfilled by using the digital receiver concept [116], sometimes referred to as software-defined radio [117]. The received microwave signal is amplified by a wideband analog receiver, downconverted to intermediate frequency and sampled by a high-speed analog/digital converter. Further filtering and quadrature detection are performed digitally, which guarantees identical receiving channels frequency responses. The output signal is a complex digital signal shifted to the baseband, sometimes referred to as a complex envelope. Nevertheless, to simplify the analysis, all further equations will be written in a continuous time domain, but discrete time equations lead to the same results.

Let us consider the case when the target can be treated as a point emitter. Such an assumption can be valid if the range extent of the target is much smaller than the range resolution cell size[1] $\Delta r = \frac{c}{B}$. To achieve a significant correlation between the received signals, it is required that the thermal emission

1. In active radar the signal propagates from the radar to the target and back to the radar. In the described passive system the signal travels one way from the target to the receiving antenna, so the range resolution is half the quality of an active radar in the same bandwidth.

from a target surface illuminates simultaneously both receiving antennas. It is equivalent to the requirement that both antennas "observe" the same part of the target's body.

The ground clutter can be divided into two classes of components. The first class is components that simultaneously illuminate both antennas, and the second class is components that illuminate only one receiver. To simplify the problem, only one point-like moving target and one point-like ground clutter source will be considered in the set of equations (7.5) and (7.6), which describe the signals received by both sensors (complex envelope of high-frequency signals):

$$x_1(t) = \xi_1(t) + a_1\xi_t\left(t - \frac{r_{t_1}}{c}\right)\exp(-2\pi j f_{d1} t) + a_{g1}\xi_g\left(t - \frac{r_{g1}}{c}\right) + \xi_{1e}(t) \tag{7.5}$$

$$x_2(t) = \xi_2(t) + a_2\xi_t\left(t - \frac{r_{t_2}}{c}\right)\exp(-2\pi j f_{d2}) + a_{g2}\xi_g\left(t - \frac{r_{g_2}}{c}\right) + \xi_{2e}(t) \tag{7.6}$$

where $\xi_1(t)$, $\xi_2(t)$ are the receiver's thermal noise components, $\xi_t(t)$ is the normalized thermal noise component generated by the target, $\xi_g(t)$ is the normalized thermal noise generated by the ground clutter, a_1, a_2 are amplitudes of the received target signals, r_{t_1}, r_{t_2} are the distances between the receiving sites and the moving target, $f_{d1} = \frac{F}{c}\frac{dr_{t_1}}{dt}, f_{d2} = \frac{F}{c}\frac{dr_{t_2}}{dt}$ are Doppler frequency shifts of the received signals originating from the target, r_{g_1}, r_{g_2} are the distances between reviving sites and the ground clutter source, a_{g1}, a_{g2} are amplitudes of ground clutter, and $\xi_{1e}(t)$, $\xi_{2e}(t)$ are noise components originating from the ground clutter or moving targets that illuminate only one sensor.

In classical radiometry the signal cross-correlation function is often exploited. In the presented case the target signals are shifted both in time and in frequency. The optimal detection (in the mean square sense) for the single target scenario (without the ground clutter) can be performed by calculating the correlation coefficient between the signal received at the first site and the signal received at the second site, appropriately shifted in time and frequency for compensation of the range and Doppler shifts. In practice, the target range and Doppler parameters are unknown, and it is necessary to repeat the procedure described above for all range and Doppler resolution cells. As a result, the 2-D range-Doppler cross-correlation function has to be calculated [118]:

$$R(r, f_d) = \int_{t=t_0}^{t_0+t_i} x_1(t) x_2^*\left(t - \frac{r}{c}\right) \exp\left(-2\pi j f_d t\right) dt. \tag{7.7}$$

A similar approach has also been described in [119] for the remote sensing of hydrometeors. The range and Doppler coordinates that maximize the cross-correlation function (7.7) are the estimates of the target range difference $(r_{t_1} - r_{t_2})$ and Doppler frequency difference $(f_{d1} - f_{d2})$. The target is considered as a moving one if the estimated Doppler frequency difference is higher than the expected Doppler spread of a stationary target. The bistatic detection range can be calculated using the following formula [115]:

$$r_{t_1} r_{t_2} = \frac{S_o T_t G_r \sqrt{t_i B}}{2\pi T_R D_0 L} \tag{7.8}$$

which is very similar to (7.3) describing the classical radiometric detection range.

The advantage of this method is the possibility of distinguishing between stationary and moving targets. A simulation result of a range-Doppler plane calculated for each range and Doppler resolution cell directly from (7.7) for a single moving target (without ground clutter) is presented in Figure 7.2. The range difference from the moving target to the sensors is equal to 6.2 km, the radial velocity difference is equal to 61 m/s, the signal received from the target is 8 dB smaller than the receiver's system noise, and the time-bandwidth product is 10,000.

A single pair of sensors can be used to detect the target and to estimate the following: the two angles to this target, the difference of ranges between the target and the sensors, and the difference of radial velocities towards the sensors. In many cases this information is sufficient for the localization and tracking of the target. A better positioning accuracy can be obtained by using several pairs of sensors. In this case the angular information is redundant and the target position (on a 2-D plane or in a 3-D space) can be calculated using only the range data.

7.3 Limitations of System Performance

The detection range determined by (7.8) can be extended by increasing integration time t_i, system bandwidth B, receiver antenna gain G, and the decreasing

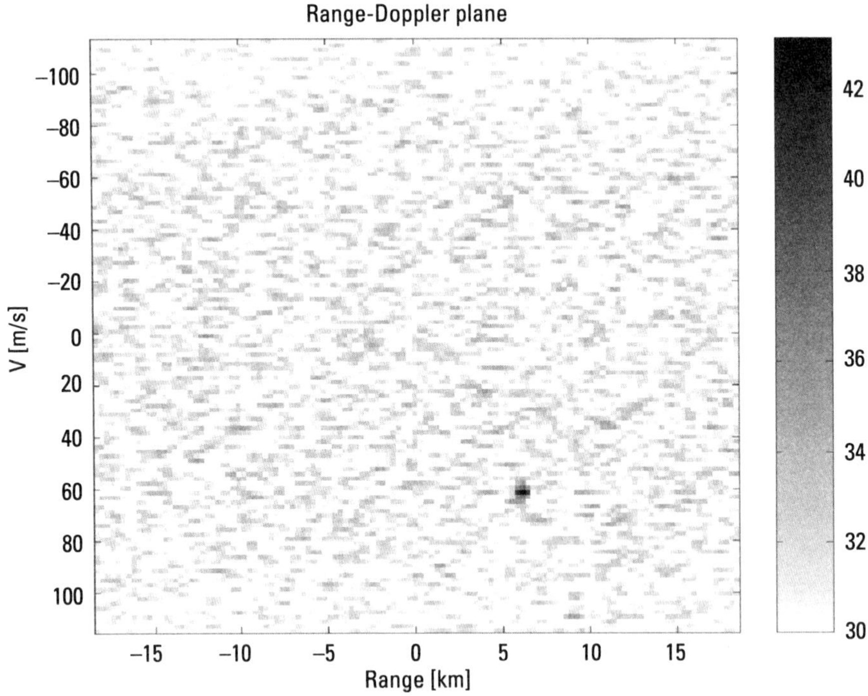

Figure 7.2 Range-Doppler correlation plane for a single moving target (grayscale is in decibels, the x-axis is the range difference between target and sensors, and the y-axis is the radial velocity difference).

of the receiver's system equivalent temperature (e.g., by the use of a cooled amplifier). Performance of the system is limited by several factors. The first is the presence of the ground clutter. The part of the signal that originated from ground clutter that illuminates both receivers introduces additional correlation peaks at the zero frequency to the correlation plane (7.7), and sidelobes at nonzero frequencies. The part of the ground clutter that illuminates only one sensor increases the overall receiver noise level as well as the level of sidelobes in the correlation plane. The main lobe component may be easily removed using various types of MTI or MTD processing [1]. The sidelobe components can be removed in an active noise radar if a reference signal is available. In radiometric systems, however, it is not possible to create such a signal, and the heavy ground clutter can reduce the detection range. The reduction of the detection range occurs when the ground clutter power is higher than the thermal noise of one of the receivers. This effect can be incorporated into (7.8) by substituting the receiver system losses L with the value:

$$L_C = \frac{kT_r B + P_g / L}{kT_r B} \tag{7.9}$$

where P_g is the total power originated from ground clutter.

Figures 7.3 and 7.4 show two examples of the range-Doppler plots obtained in a simulation where both a moving target and the ground clutter are present. The first one illustrates the case where the ground clutter power is smaller than the thermal noise of one of the receivers. The scenario consists of one moving target and four point-like ground clutter sources. The signal received from the moving targets is 8 dB smaller than the system thermal noise, and the signal received from each clutter source is 12 dB smaller than the system noise. The growth of the sidelobe level is no more than 1 dB in comparison with the clutter free environment.

In the scenario presented in Figure 7.4, the ground clutter power is much greater than the power of the receiver's thermal noise. The scenario consists of one moving target and two point-like ground clutter sources (generating a signal 8 dB higher than the system noise). The sidelobe level increases significantly, and in this case a small moving target cannot be detected. The sidelobes

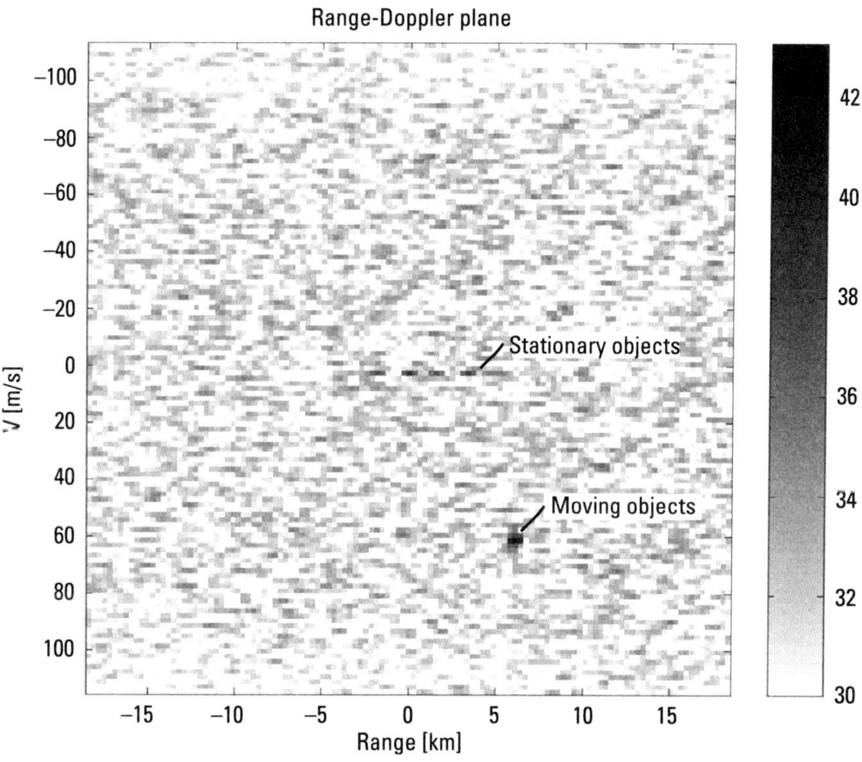

Figure 7.3 Range-Doppler correlation plane, a single moving target and four weak ground clutter sources (grayscale is in decibels, the *x*-axis is the range difference between the target and sensors, and the *y*-axis is the radial velocity difference).

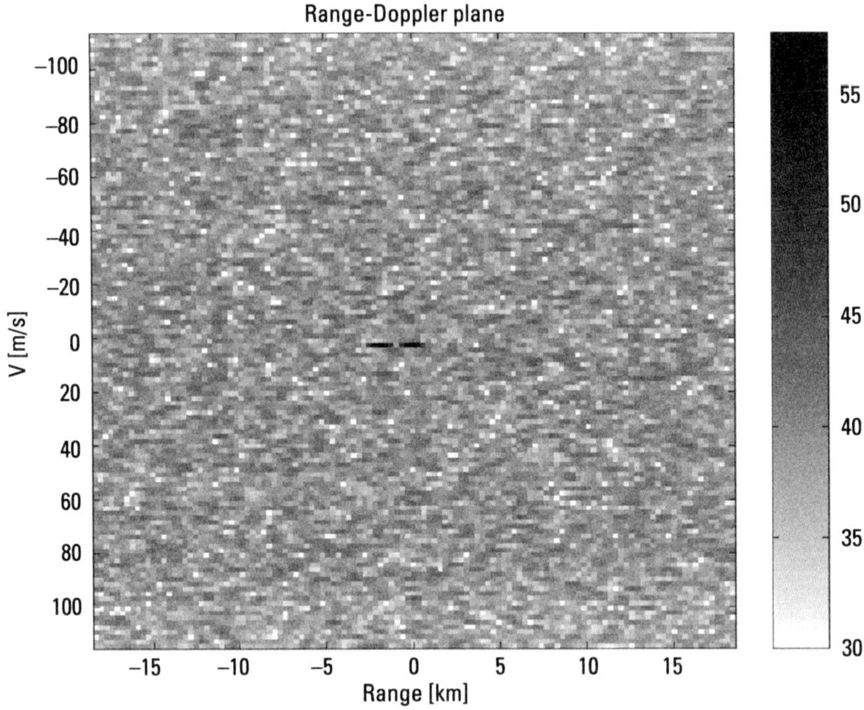

Figure 7.4 Range-Doppler correlation plane, a single moving target and two strong ground clutter sources (grayscale is in decibels, the *x*-axis is the range difference between the target and sensors, and the *y*-axis is the radial velocity difference). The moving target is masked by ground clutter sidelobes.

generated by the ground clutter can be reduced by increasing the time-bandwidth product and by decreasing the antenna beamwidth.

The second limitation is caused by the target range migration. The range resolution of the system is equal to $\frac{c}{B}$. The integration time should be limited by the time interval when a target remains within the resolution cell. This limitation is described by the inequality:

$$t_i < \frac{c}{Bv} \tag{7.10}$$

where v is the difference of the object range velocities towards both receivers. Extension of the integration time can be achieved by adding additional terms in the cross-correlation function (7.7) correcting the range migration of the target [118]. Assuming the constant target velocity v, the range can be expressed as $r = r_0 + vt$, and the correlation equation can be rewritten in the form:

$$R(r_0,V)=\int_{t=t_0}^{t_0+t_i} x_1(t)x_2^*\left(t-\frac{r_0+vt}{c}\right)\exp\left(-2\pi j\frac{vF}{c}t\right)dt \qquad (7.11)$$

The correlation described by (7.11) can be calculated effectively using the stretch processing concept described in Chapter 3.

The third limitation is caused by the target Doppler frequency migration. The Doppler frequency resolution of the system is equal to $\frac{1}{t_i}$, and the velocity resolution is $\frac{c}{t_iF}$. For maneuvering targets with the constant acceleration a, the integration time is limited by the inequality:

$$t_i<\sqrt{\frac{c}{aF}}=\sqrt{\frac{\lambda}{a}} \qquad (7.12)$$

A further extension of integration time can be obtained by incorporating the acceleration term into the correlation function (7.11):

$$\begin{aligned}R(r_0,v,a)=&\int_{t=t_0}^{t_0+t_i} x_1(t)x_2^*\left(t-\frac{r_0+vt+at^2/2}{c}\right)\\&\exp\left(-2\pi j\frac{(v+at/2)F}{c}t\right)dt\end{aligned} \qquad (7.13)$$

It is possible to also take into account higher-order terms (e.g., acceleration change), but this would lead to very high computational complexity, which could not be implemented nowadays in real time. Another possibility to overcome the presented limitations is to combine the coherent integration with the noncoherent one. A short coherent integration time leads to a wider velocity resolution cell. To overcome the acceleration problem, a long data record can be divided into several shorter subblocks, and for each subblock the coherent integration can be performed. The inter-subblock noncoherent integration will decrease the system sensitivity in comparison to single subblock sensitivity by only a few decibels, and the numerical complexity will be much lower. A similar approach can be used in the case where limitation is caused by the range migration of the target. In this case it is possible to expand the range resolution cell by dividing the receiver bandwidth into several subbands [120]. The coherent integration can be performed separately for each subband and noncoherent integration can be applied to combine the data from all subbands. It is also possible to combine these two methods to achieve a high time-bandwidth product and high sensitivity, only a few decibels lower than using the compound coherent integration.

7.4 Passive Target Imaging

In a modern radar system the detection of a moving target is only the first step in data signal processing. The next step is to identify the target. Target identification may be performed using different methods. The classical secondary surveillance radar (SSR) or identification friend or foe (IFF) uses active identification. Such a radar sends an identification request to the target, and the target responds sending an identification code to the radar. Such an identification method may be achieved only for cooperating targets. For noncooperative targets identification may be done using high-resolution target imaging. For this purpose optical, infrared, or radar imaging techniques are often applied. Optical or infrared imaging is based on high angular resolution of the optical systems, while radar imaging is based on inverse synthetic aperture radar (ISAR) technology exploiting the coherent integration of target echoes.

The coherent integration technique, similar to the ISAR technique, can be used to obtain the high-resolution target image using the target's thermal emission.

Let us consider a target composed of separated emission point-like centers both in the cross-track dimension and the along track dimension. The target is moving parallel to the Y-axis with a velocity of v. Two radiometric sensors observe the target in the time interval $(-t_i/2..t_i/2)$. The sensors are placed in positions $(0,\ Y_0)$, $(0, -Y_0)$, respectively.

An example scenario is presented in Figure 7.5. The distances between the target and sensors are equal to:

$$\begin{aligned} r_1(t) &= \sqrt{X_0^2 + (vt - Y_0)^2} \\ r_2(t) &= \sqrt{X_0^2 + (vt + Y_0)^2} \end{aligned} \tag{7.14}$$

The signals received by both sensors can be described by the equations:

$$x_1(t) = \xi_1(t) + a_1 \xi_t \left(t - \frac{r_1}{c} \right) \exp\left(-2\pi j f_{d1} t\right) \tag{7.15}$$

$$x_2(t) = \xi_2(t) + a_2 \xi_t \left(t - \frac{r_2}{c} \right) \exp(-2\pi j f_{d2} t) \tag{7.16}$$

For a short integration time optimal detection processing can be described by (7.7). For a longer integration time it is necessary to modify the detection equation and take into consideration the range migration effect and relation between the Doppler frequency and the range to the target. The range difference between a point-like emitter and receiving sites can be approximated by

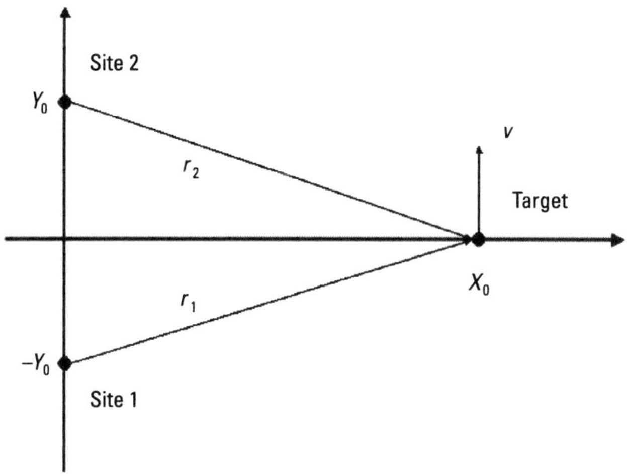

Figure 7.5 An example of a radiometric imaging scenario.

the Taylor series. Taking into account the first derivation, one can obtain the formula:

$$\Delta r(t) = r_1(t) - r_2(t) \approx \frac{2Y_0 v}{\sqrt{X_0^2 + Y_0^2}} t \tag{7.17}$$

The Doppler frequency difference between the received signals can be approximated by the following formula:

$$f_d(t) \approx \frac{2Y_0 v}{\lambda\sqrt{X_0^2 + Y_0^2}} \tag{7.18}$$

It is easy to notice that the Doppler frequency depends on the range to the target X_0 and the distance between the receiving sites $2Y_0$. The optimal correlation schema can then be expressed as:

$$\begin{gathered} R(X_0, v) = \int_{t=-t_i/2_0}^{t_i/2} x_1(t) x_2^*\left(t - \frac{\Delta r(t)}{c}\right) \exp(-2\pi j f_d t) dt \approx \\ \int_{t=-t_i/2_0}^{t_i/2} x_1(t) x_2^*\left(t - \frac{2Y_0 vt}{c\sqrt{X_0^2 + Y_0^2}}\right) \exp\left(-2\pi j \frac{2Y_0 v}{\lambda\sqrt{X_0^2 + Y_0^2}} t\right) dt \end{gathered} \tag{7.19}$$

If the target is a complex one and can be treated as a set of emission centers placed at different ranges, then each emission center contribution will have a different Doppler frequency, and those centers can be distinguished if their

Doppler frequencies differ more than the frequency resolution. The Doppler frequency resolution achieved using correlation equation (7.19) is equal to $1/T$, and the final range resolution of the target can be expressed by the formula:

$$\Delta x \approx \frac{\lambda\left(X_0^2 + Y_0^2\right)^{3/2}}{2Y_0X_0vt_i} \tag{7.20}$$

An example of a target range resolution for a radiometric system working at 20 GHz with 1-second integration time and 1 km of distance between the receiving sites versus the range to the target (X_0) for $v = 100$ m/s, is presented in Figure 7.6. The resolution at the distance of 1 km is much better than 1m, for 10 km it is equal to 10m and for 100 km it is equal to 1 km. The resolutions for long ranges are not too good, but can be improved using a longer integration time and a longer baseline. A similar analysis may be carried out to calculate the along-track resolution, but in this case the resolution is proportional to the receiver's bandwidth:

$$\Delta y \approx \frac{\left(X_0^2 + Y_0^2\right)^{1/2} c}{2Y_0B} \tag{7.21}$$

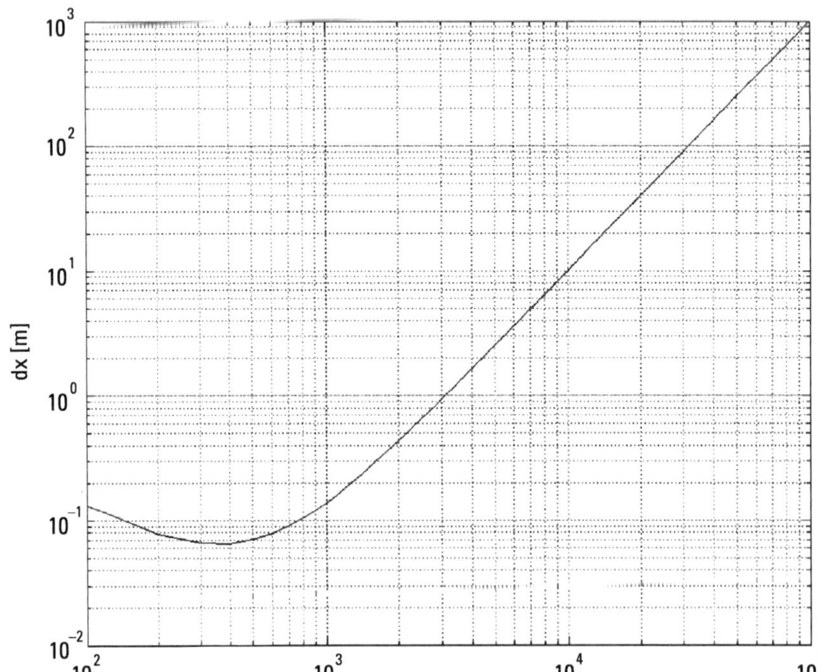

Figure 7.6 Target range resolution in the imaging mode.

In the scenario presented the simplest possible configuration was analyzed, but it is not difficult at all to extend the analysis to some more complicated scenarios.

As presented above, the passive noise system can be used both for moving target detection, tracking and target imaging. The additional limitation of the passive detection and imaging methods presented comes from the existing transmitter network. At present, it is very difficult to find a frequency band free of any man-made emission, and each man-made emission contributes to the overall system noise level and causes the decrease of passive system sensitivity. The influence of man-made emissions can be decreased by the adaptive cancellation of unwanted emissions [121] using additional reference antennas.

As seen in this chapter, it is possible to use the radiometric coherent processing concept to detect moving targets and also to image the target. For this purpose it is necessary to build a set of receiving stations tuned to the same frequency and send all received signals to the processing center. To have sufficient sensitivity, it is necessary to find the frequency band free of interferences, which can be more and more difficult in the present time as an increasing amount of frequency bands are allocated for communication purposes.

References

[1] Skolnik, M. I., *Radar Handbook*, 3rd ed., New York: McGraw-Hill, 2008.

[2] Prabaswara, A., A. Munir, and A. B. Suksmono, "GNU Radio Based Software-Defined FMCW Radar for Weather Surveillance Application," *2011 6th International Conference on Telecommunication Systems, Services, and Applications (TSSA)*, October 20–21, 2011, pp. 227–230.

[3] Tessmann, A., et al., "Compact Single-Chip W-Band FMCW Radar Modules for Commercial High-Resolution Sensor Applications," *IEEE Transactions on Microwave Theory and Techniques*, Vol. 50, No. 12, December 2002, pp. 2995–3001.

[4] Yamaguchi, J., et al., "Sensitivity of FMCW 95GHz Cloud Radar for High Clouds," *2006 Asia-Pacific Microwave Conference (APMC 2006)*, December 12–15, 2006, pp. 1841–1846.

[5] Im, Y. -T., J. -H. Lee, and S. -O. Park, "A Pulse-Doppler and FMCW Radar Signal Processor for Surveillance," *2011 3rd International Asia-Pacific Conference on Synthetic Aperture Radar (APSAR)*, September 26–30, 2011, pp. 1–4.

[6] Wojtkiewicz, A., et al., "Use of Polynomial Phase Modeling to FMCW Radar. Part C: Estimation of Target Acceleration in FMCW Radars," *NATO Research and Technology Agency, Sensors & Electronics Technology Symposium on Passive and LPI (Low Probability of Intercept) Radio Frequency Sensors*, Warsaw, Poland, April 23–25, 2001, paper #40C.

[7] Pace, P. F., *Detecting and Classifying Low Probability of Intercept Radar*, Norwood, MA: Artech House, 2003.

[8] Komarov, I. V., S. M. Smolskiy, and D. K. Barton, *Fundamentals of Short-Range FM Radar*, Norwood, MA: Artech House, 2003.

[9] Lukin K., "Radar Design Using Noise/Random Waveforms," *2006 International Radar Symposium*, Krakow, Poland, May 24–26, 2006, pp. 355–358.

[10] Lukin, K., "Noise Radar Technology: The Principles and Short Overview," *Applied Radio Electronics*, Vol. 4, No. 1, 2005, pp. 4–11.

[11] Lukin, K. A., "The Principles of Noise Radar Technology," *The First International Workshop on the Noise Radar Technology*, Crimea, September 18–22, 2002, pp. 13–22.

[12] Byrnes, J., (ed.), *Advances in Sensing with Security Applications* (NATO Security Through Science Series, New York: Springer, 2006.

[13] Kulpa, K., et al., "Signal Processing in Noise Radar Technology [Editorial]," *IET Radar, Sonar & Navigation*, Vol. 2, No. 4, August 2008, pp. 229–232.

[14] Greco, M., F. Gini, and A. Farina, "Radar Detection and Classification of Jamming Signals Belonging to a Cone Class," *IEEE Transactions on Signal Processing*, Vol. 56, No. 5, May 2008, pp. 1984–1993.

[15] Rao, M. S., "EMI/EMC Effects on EW Receiver Systems of Military Aircraft," *10th International Conference on Electromagnetic Interference & Compatibility (INCEMIC 2008)*, November 26–27, 2008, pp. 63–67.

[16] Nan, L., and Q. Changwen, "Research on Jamming Synthetic Aperture Radar Technologies," *1st Asian and Pacific Conference on Synthetic Aperture Radar (APSAR 2007)*, November 5–9, 2007, pp. 563–566.

[17] Yanbin, Z., "Technology of Smart Noise Jamming Based on Multiplication Modulation," *2011 International Conference on Electric Information and Control Engineering (ICEICE)*, April 15–17, 2011, pp. 4557–4559.

[18] Jun-jun, S., B. Da-Ping, and X. Lei, "Novel Evaluation Method of Jamming Effect on ISAR Based on Target Detection," *2nd Asian-Pacific Conference on Synthetic Aperture Radar (APSAR 2009)*, October 26–30, 2009, pp. 892–895.

[19] Yong, Y., W. -M. Zhang, and J. -H. Yang, "Study on Frequency-Shifting Jamming to Linear Frequency Modulation Pulse Compression Radars," *2009 International Conference on Wireless Communications & Signal Processing (WCSP 2009)*, November 13–15, 2009, pp. 1–5.

[20] Zhang, L., J. Ren, and and T. Li, "Time-Varying Jamming Modeling and Classification," *IEEE Transactions on Signal Processing*, Vol. 60, No. 7, July 2012, pp. 3902–3907.

[21] Griffiths, H. D., et al., "Measurement and Analysis of Ambiguity Functions of Off-Air Signals for Passive Coherent Location," *Electronics Letters*, Vol. 39, No. 13, June 26, 2003, pp. 1005–1007.

[22] Zoeller, C. L., M. C. Budge, and M. L. Moody, Jr., "Passive Coherent Location Radar Demonstration," *Proceedings of the Thirty-Fourth Southeastern Symposium on System Theory*, March 18–19, 2002, pp. 358–362.

[23] Oyedokun, T., "Feasibility Study of DTV Based PCL Radar in South Africa," *2011 European Radar Conference (EuRAD)*, October 12–14, 2011, pp. 186–189.

[24] Dawidowicz, B., and K. S. Kulpa, "Experimental Results from PCL Radar on Moving Platform," *2008 International Radar Symposium*, May 21–23, 2008, pp. 1–4.

[25] Mousavi, M. R., A. Jafargholi, and M. M. Nayebi, "Fast and Accurate Method for PCL Radar Detection in Noisy Environment," *3rd European Radar Conference 2006 (EuRAD 2006)*, September 13–15, 2006, pp. 33–36.

[26] Stinco, P., et al., "Sensor Selection in PCL Radar Systems Based on Bistatic PCRLB," *2012 Tyrrhenian Workshop on Advances in Radar and Remote Sensing (TyWRRS)*, September 12–14, 2012, pp. 41–45.

[27] Malanowski, M., K. Kulpa, and J. Misiurewicz, "PaRaDe: PAssive RAdar DEmonstrator Family Development at Warsaw University of Technology," *2008 Microwaves, Radar and Remote Sensing Symposium (MRRS 2008)*, September 22–24, 2008, pp. 75–78.

[28] Howland, P. E., "Target Tracking Using Television-Based Bistatic Radar," *IEE Proceedings Radar, Sonar and Navigation*, Vol. 146, No. 3, June 1999, pp. 166–174.

[29] Howland, P. E., D. Maksimiuk, and G. Reitsma, "FM Radio Based Bistatic Radar," *IEE Proceedings Radar, Sonar and Navigation*, Vol. 152, No. 3, June 3, 2005, pp. 107–115.

[30] Matthes, D., "Convergence of ESM Sensors and Passive Covert Radar," *2005 IEEE International Radar Conference*, May 9–12, 2005, pp. 430–444.

[31] Di Lallo, A., R. Fulcoli, and L. Timmoneri, "Adaptive Spatial Processing Applied to a Prototype Passive Covert Radar: Test with Real Data," *2008 International Conference on Radar*, September 2–5, 2008, pp. 139–143.

[32] Kulpa, K. S., and J. Misiurewicz, "Stretch Processing for Long Integration Time Passive Covert Radar," *2006 International Conference on Radar (CIE '06)*, October 16–19, 2006, pp. 1–4.

[33] Coman, C. I., S. Gelsema, and C. Stolk, "Array Antennas for Passive Covert Radar Applications," *2010 Proceedings of the Fourth European Conference on Antennas and Propagation (EuCAP)*, April 12–16, 2010, pp. 1–4.

[34] Di Lallo, A., et al., "Design, Development and Test on Real Data of an FM Based Prototypical Passive Radar," *2008 IEEE Radar Conference (RADAR '08)*, May 26–30, 2008, pp. 1–6.

[35] Sahr, J. D., and F. D. Lind, "The Manastash Ridge Radar: A Passive Bistatic Radar for Upper Atmospheric Radio Science," *Radio Sci.*, Vol. 32, No. 6, 1997, pp. 2345–2358.

[36] Bongioanni, C., F. Colone, and P. Lombardo, "Performance Analysis of a Multi-Frequency FM Based Passive Bistatic Radar," *2008 IEEE Radar Conference (RADAR '08)*, May 26–30, 2008, pp. 1–6.

[37] Li, S., et al., "Detection of Moving Target in FM Broadcast-Based Passive Radar," *2007 International Conference on Wireless Communications, Networking and Mobile Computing, (WiCom 2007)*, September 21–25, 2007, pp. 649–652.

[38] Tasdelen, A. S., and H. Koymen, "Range Resolution Improvement in Passive Coherent Location Radar Systems Using Multiple FM Radio Channels," *The Institution of Engineering and Technology Forum on Waveform Diversity and Design in Communications, Radar and Sonar*, November 22, 2006, pp. 23–31.

[39] O'Hagan, D. W., and C. J. Baker, "Passive Bistatic Radar (PBR) Using FM Radio Illuminators of Opportunity," *New Trends for Environmental Monitoring Using Passive Systems*, October 14–17, 2008, pp. 1–6.

[40] Kulpa, K., et al., "On-Board PCL Systems for Airborne Platform Protection," *2011 Tyrrhenian International Workshop on Digital Communications/Enhanced Surveillance of Aircraft and Vehicles (TIWDC/ESAV)*, September 12–14, 2011, pp. 119–122.

[41] Howland, P. E., "A Passive Metric Radar Using the Transmitters of Opportunity," *Int. Conf.on Radar*, Paris, France, May 1994, pp. 251–256.

[42] Nordwall, B. D., "'Silent Sentry' A New Type of Radar," *Aviation Week & Space Technology*, No. 30, 1998, pp. 70–71.

[43] Howland, P. E., "Target Tracking Using Television-Based Bistatic Radar," *IEE Proc. Radar, Sonar & Navig.*, Vol. 146, No. 3, June 1999.

[44] Sun, L. -X., J. Liu, and S. -X. Chen, "Analogue TV Signal-Based Passive Radar Onboard Single Air-Platform," *2006 International Conference on Radar (CIE '06)*, October 16–19, 2006, pp. 1–4.

[45] Wang, H., J. Wang, and L. Zhong, "Mismatched Filter for Analogue TV-Based Passive Bistatic Radar," *IET Radar, Sonar & Navigation*, Vol. 5, No. 5, June 2011, pp. 573–581.

[46] Griffiths, H. D., and C. J. Baker, "Measurement and Analysis of Ambiguity Functions of Passive Radar Transmissions," *2005 IEEE International Radar Conference*, May 9–12, 2005, pp. 321–325.

[47] Griffiths, H. D., et al., "Bistatic Radar Using Satellite-Borne Illuminators of Opportunity," *Proc. International Conference RADAR 2002*, October 2002, pp. 1–5.

[48] Marques, P., et al., "A Pedagogical Passive RADAR Using DVB-S Signals," *2011 3rd International Asia-Pacific Conference on Synthetic Aperture Radar (APSAR)*, September 26–30, 2011, pp. 1–4.

[49] Cristallini, D., et al., "Space-Based Passive Radar Enabled by the New Generation of Geostationary Broadcast Satellites," *2010 IEEE Aerospace Conference*, March 6–13, 2010, pp. 1–11.

[50] Dickinson, D. F., and W. C. Straka, "Tethered Satellite Antenna Arrays for Passive Radar Systems," *1990 IEEE Aerospace Applications Conference Digest*, February 4–9, 1990, pp. 117–125.

[51] Pengge, M., et al., "Navigation Satellite Passive Radar Moving Target Detection and SAR Imaging Based on FRFT," *2010 First International Conference on Pervasive Computing Signal Processing and Applications (PCSPA)*, September 17–19, 2010, pp. 767–770.

[52] Wang, J., H. Shunji, and C. Fengyun, "Application of GPS to Exospheric Target Detection Using Missile-Borne Passive Radar," *14th IEEE Proceedings on Personal, Indoor and Mobile Radio Communications 2003 (PIMRC 2003)*, Vol. 3, September 7–10, 2003, pp. 2226–2229.

[53] Xu, D. -J., et al., "A Weak Target Detection Method for GPS Based Passive Radar," *2011 3rd International Conference on Advanced Computer Control (ICACC)*, January 18–20, 2011, pp. 278–281.

[54] Maussang, F., et al., "GPS Passive Bistatic Radar System in Oceanic Environment: Detection Performance Estimation," *2011 IEEE OCEANS*, Spain, June 6–9, 2011, pp. 1–5.

[55] Mojarrabi, B., et al., "Power Budget Study for Passive Target Detection and Imaging Using Secondary Applications of GPS Signals in Bistatic Radar Systems," *2002 IEEE International Geoscience and Remote Sensing Symposium (IGARSS '02)*, Vol. 1, 2002, pp. 449–451.

[56] Suberviola, I., I. Mayordomo, and J. Mendizabal, "Experimental Results of Air Target Detection with a GPS Forward-Scattering Radar," *IEEE Geoscience and Remote Sensing Letters*, Vol. 9, No. 1, January 2012, pp. 47–51.

[57] Behar, V., and C. Kabakchiev, "Detectability of Air Targets Using Bistatic Radar Based on GPS L5 Signals," *2011 Proceedings International Radar Symposium (IRS)*, September 7–9, 2011, pp. 212–217.

[58] Capria, A., et al., "DVB-T Passive Radar for Vehicles Detection in Urban Environment," *2010 IEEE International Geoscience and Remote Sensing Symposium (IGARSS)*, July 25–30, 2010, pp. 3917–3920.

[59] Conti, M., et al., "High Range Resolution Multichannel DVB-T Passive Radar," *IEEE Aerospace and Electronic Systems Magazine*, Vol. 27, No. 10, October 2012, pp. 37–42.

[60] Palmer, J. E., et al., "DVB-T Passive Radar Signal Processing," *IEEE Transactions on Signal Processing*, Vol. PP, No. 99, 2012.

[61] Gao, Z., et al., "DVB-T Signal Cross-Ambiguity Functions Improvement for Passive Radar," *2006 International Conference on Radar (CIE '06)*, October 16–19, 2006, pp. 1–4.

[62] Kuschel, H., et al., "On the Resolution Performance of Passive Radar Using DVB-T Illuminations," *2010 11th International Radar Symposium (IRS)*, June 16–18, 2010, pp. 1–4.

[63] Malanowski, M., K. Kulpa, and K. E. Olsen, "Extending the Integration Time in DVB-T-Based Passive Radar," *2011 European Radar Conference (EuRAD)*, October 12–14, 2011, pp. 190–193.

[64] Harms, H. A., L. M. Davis, and J. Palmer, "Understanding the Signal Structure in DVB-T Signals for Passive Radar Detection," *2010 IEEE Radar Conference*, May 10–14, 2010, pp. 532–537.

[65] Olsen, K. E., and K. Woodbridge, "Multiband Passive Bistatic DVB-T Radar Range Resolution Improvements and Implications," *2012 13th International Radar Symposium (IRS)*, May 23–25, 2012, pp. 28–31.

[66] Raout, J., "Sea Target Detection Using Passive DVB-T Based Radar," *2008 International Conference on Radar*, September 2–5, 2008, pp. 695–700.

[67] Coleman, C. J., R. A. Watson, and H. Yardley, "A Practical Bistatic Passive Radar System for Use with DAB and DRM Illuminators," *2008 IEEE Radar Conference (RADAR '08)*, May 26–30, 2008, pp. 1–6.

[68] Yardley, H. J., "Bistatic Radar Based on DAB Illuminators: The Evolution of a Practical System," *IEEE Aerospace and Electronic Systems Magazine*, Vol. 22, No. 11, November 2007, pp. 13–16.

[69] Poullin, D., "Passive Detection Using Digital Broadcasters (DAB, DVB) with COFDM Modulation," *IEE Proceedings Radar, Sonar and Navigation*, Vol. 152, No. 3, June 3, 2005, pp. 143–152.

[70] Poullin, D., and M. Flecheux, "Recent Progress in Passive Coherent Location (PCL) Concepts and Technique in France Using DAB or FM Broadcasters," *2008 IEEE Radar Conference (RADAR '08)*, May 26–30, 2008, pp. 1–5.

[71] Malanowski, M., et al., "Silent Police Radar," *Proc. of 3rd International Workshop on Inteligent Transportation (WIT-2006)*, Hamburg, Germany, March 14–15, 2006, pp. 117–122.

[72] Krysik, P., et al., "Detection of Fast Maneuvering Air Targets Using GSM Based Passive Radar," *2012 13th International Radar Symposium (IRS)*, May 23–25, 2012, pp. 69–72.

[73] De Maio, A., et al., "Experimental Verification of a Two-State Model for the Cumulative Distribution Function of GSM Passive Radar Clutter," *2008 IEEE Instrumentation and Measurement Technology Conference Proceedings (IMTC 2008)*, May 12–15, 2008, pp. 1288–1293.

[74] Zemmari, R., M. Daun, and U. Nickel, "Maritime Surveillance Using GSM Passive Radar," *2012 13th International Radar Symposium (IRS)*, May 23–25, 2012, pp. 76–82.

[75] Yunfu, Y., et al., "Direct Decomposing Modulation Phase in GSM Passive Radar," *2007 International Symposium on Microwave, Antenna, Propagation and EMC Technologies for Wireless Communications*, August 16–17, 2007, pp. 1412–1415.

[76] Krysik, P., et al., "Velocity Measurement and Traffic Monitoring Using a GSM Passive Radar Demonstrator," *IEEE Aerospace and Electronic Systems Magazine*, Vol. 27, No. 10, October 2012, pp. 43–51.

[77] Tan, D. K. P., et al., "GSM Based Through-the-Wall Passive Radar Demonstrator for Motion Sensing," *New Trends for Environmental Monitoring Using Passive Systems*, October 14–17, 2008, pp. 1–4.

[78] Mendi, U. M. D., and B. K. Sarkar, "Passive Radar Using Multiple GSM Transmitting Stations," *2006 International Radar Symposium (IRS 2006)*, May 24–26, 2006, pp. 1–4.

[79] De Maio, A., et al., "Measurement and Comparative Analysis of Clutter for GSM and UMTS Passive Radars," *IET Radar, Sonar & Navigation*, Vol. 4, No. 3, June 2010, pp. 412–423.

[80] Neyt, X., et al., "Feasibility of STAP for Passive GSM-Based Radar," *2006 IEEE Conference on Radar*, April 24–27, 2006, p. 6.

[81] Wang, Q., C. Hou, and Y. Lu, "An Experimental Study of WiMAX-Based Passive Radar," *IEEE Transactions on Microwave Theory and Techniques*, Vol. 58, No. 12, December 2010, pp. 3502–3510.

[82] Gao, G., Q. Wang, and C. Hou, "Power Budget and Performance Prediction for WiMAX Based Passive Radar," *2011 6th International Conference on Pervasive Computing and Applications (ICPCA)*, October 26–28, 2011, pp. 517–520.

[83] Chetty, K., et al., "Passive Bistatic WiMAX Radar for Marine Surveillance," *2010 IEEE Radar Conference*, May 10–14, 2010, pp. 188–193.

[84] Colone, F., P. Falcone, and P. Lombardo, "Ambiguity Function Analysis of WiMAX Transmissions for Passive Radar," *2010 IEEE Radar Conference*, May 10–14, 2010, pp. 689–694.

[85] del Arroyo, J. R. G., J. Jackson, and M. Temple, "WiMAX Ambiguity Function for PCL Systems," *Proceedings of the IEEE 2010 National Aerospace and Electronics Conference (NAECON)*, 2010, pp. 54–59.

[86] Chetty, K., G. E. Smith, and K. Woodbridge, "Through-the-Wall Sensing of Personnel Using Passive Bistatic WiFi Radar at Standoff Distances," *IEEE Transactions on Geoscience and Remote Sensing*, Vol. 50, No. 4, April 2012, pp. 1218–1226.

[87] Maechler, P., N. Felber, and H. Kaeslin, "Compressive Sensing for WiFi-Based Passive Bistatic Radar," *2012 Proceedings of the 20th European Signal Processing Conference (EUSIPCO)*, August 27–31, 2012, pp. 1444–1448.

[88] Falcone, P., et al., "Localization and Tracking of Moving Targets with WiFi-Based Passive Radar," *2012 IEEE Radar Conference (RADAR)*, May 7–11, 2012, pp. 0705–0709.

[89] Kulpa, K., et al., "Concept of Multistatic Passive Radar Based on Wireless Packet Communication Systems," *2011 IEEE CIE International Conference on Radar*, Vol. 1, October 24–27, 2011, pp. 149–152.

[90] Thompson, J. S., et al., "Generalised Algorithm for DOA Estimation in a Passive Sonar," *IEE Proceedings F Radar and Signal Processing*, Vol. 140, No. 5, October 1993, pp. 339–340.

[91] Pastorino, M., and A. Randazzo, "A Smart Antenna for the DOA Estimation of Impinging Signals and Passive Obstacle Detection for Homeland Security," *Proceedings of the 2005 IEEE International Workshop on Measurement Systems for Homeland Security, Contraband Detection and Personal Safety Workshop (IMS 2005)*, March 29–30, 2005, pp. 70–75.

[92] Wang, J., et al., "Passive Location to Moving Emitter Using DOA and Its Rate of Change," *Congress on Image and Signal Processing (CISP '08)*, Vol. 4, May 27–30, 2008, pp. 229–233.

[93] Dersan, A., and Y. Tanik, "Passive Radar Localization by Time Difference of Arrival," *Proceedings Military Communication Conference (MILCOM 2002)*, Vol. 2, October 7–10, 2002, pp. 1251–1257.

[94] Reynet, O., L. Jaulin, and G. Chabert, "Robust TDOA Passive Location Using Interval Analysis and Contractor Programming," *2009 International Radar Conference: Surveillance for a Safer World*, October 12–16, 2009, pp. 1–6.

[95] Basciftci, C. H., and H. Balaban, "Analysis of Signal Bandwith Effect on the Performance of Passive TDOA Estimation," *2011 IEEE 19th Conference on Signal Processing and Communications Applications (SIU)*, April 20–22, 2011, pp. 355–358.

[96] Liang, Q., et al., "TDoA for Passive Localization: Underwater Versus Terrestrial Environment," *IEEE Transactions on Parallel and Distributed Systems*, No. 99, pp.1, 0

[97] Mikhalev, A., E. J. Hughes, and R. F. Ormondroyd, "Comparison of Hough Transform and Particle Filter Methods of Passive Emitter Geolocation Using Fusion of TDOA and AOA Data," *2010 13th Conference on Information Fusion (FUSION)*, July 26–29, 2010, pp. 1–8.

[98] Koch, V., and R. Westphal, "New Approach to a Multistatic Passive Radar Sensor for Air/Space Defence," *IEEE Aerospace and Electronic Systems Magazine*, Vol. 10, No. 11, November 1995, pp. 24–32.

[99] Guvenel, C., et al., "An Uncooled Infrared Camera with Embedded Real-Time Target Tracking Capability," *2012 20th Signal Processing and Communications Applications Conference (SIU)*, April 18–20, 2012, pp. 1–4.

[100] Li, C., et al., "Robust Target Detection and Tracking in Outdoor Infrared Video," *2008 IEEE International Conference on Acoustics, Speech and Signal Processing (ICASSP 2008)*, March 31–April 4, 2008, pp. 1489–1492.

[101] Dastner, K., B. Kohler, and F. Opitz, "3d Object Extraction and Tracking for Infra Red and Day Light Camera Systems," *2011 Proceedings International Radar Symposium (IRS)*, September 7–9, 2011, pp. 737–742.

[102] Andreone, L., et al., "Vehicle Detection and Localization in Infra-Red Images," *Proceedings of the IEEE 5th International Conference on Intelligent Transportation Systems*, 2002, pp. 141–146.

[103] Ulaby, F., R. Moore, and A. Fung, *Microwave Remote Sensing: Active and Passive, Vol. I: Microwave Remote Sensing Fundamentals and Radiometry*, Reading, MA: Addison-Wesley, 1981.

[104] Thomson, A. R., J. M. Moran, and J. G. W. Swenson, *Interferometry and Synthesis in Radio Astronomy*, New York: John Wiley & Sons, 1986.

[105] Clapp, R., and J. Maxwell, "Complex-Correlation Radiometer," *IEEE Transactions on J. Antennas and Propagation*, [legacy, pre-1988], Vol. 15, No. 2, March 1967, pp. 286–290.

[106] Piepmeier, J. R., A. J. Gasiewski, and J. E. Almodovar, "Advances in Microwave Digital Radiometry," *Proceedings on the 2000 IEEE International Geoscience and Remote Sensing Symposium (IGARSS 2000)*, Vol. 7, July 24–28, 2000, pp. 2830–2833.

[107] Edelsohn, C. R., "Applications of Synthetic Aperture Radiometry," *1994 International Geoscience and Remote Sensing Symposium (IGARSS '94): Surface and Atmospheric Remote Sensing: Technologies, Data Analysis and Interpretation*,Vol. 3, August 8–12, 1994, pp. 1326–1328.

[108] Bardati, F., et al., "Microwave Radiometry for Medical Thermal Imaging: Theory and Experiment," *IEEE MTT-S International Microwave Symposium Digest*, Vol. 3, June 1–5, 1992, pp. 1287–1290.

[109] Brown, E. R., et al., "Wide-Band Radiometry for Remote Sensing of Oil Films on Water," *IEEE Transactions on Microwave Theory and Techniques*, Vol. 46, No. 12, December 1998, pp. 1989–1996.

[110] Wilkinson, A. J., and M. R. Inggs, "Radiometry for Landmine Detection," *Proceedings of the 1998 South African Symposium on Communications and Signal Processing (COMSIG '98)*, September 7–8, 1998, pp. 477–482.

[111] Sonnenschein, A., and P. Fishman, "Radiometric Detection of Spread-Spectrum Signals in Noise of Uncertain Power," *IEEE Transactions on Aerospace and Electronic Systems*, Vol. 28, No. 3, July 1992, pp. 654–660.

[112] Camps, A. J., and C. T. Swift, "A Two-Dimensional Doppler-Radiometer for Earth Observation," *IEEE Transactions on Geoscience and Remote Sensing*, Vol. 39, No. 7, July 2001.

[113] Sousa, M., and J. Betz, "Limitation of Radiometer Performance in Spherically Invariant Noise," *IEEE Transactions on Aerospace and Electronic Systems*, Vol. 31, No. 1, January 1992, pp. 65–68.

[114] Mills, R. F., and G. E. Prescott, "A Comparison of Various Radiometer Detection Models," *IEEE Transactions on Aerospace and Electronic Systems*, Vol. 32, No. 1, January 1996, pp. 467–473.

[115] Kulpa, K. S., "Passive Multi-Static Radiometric Detection of Moving Targets," *Proc. XXV International Conference. on Microwaves, Radar and Wireless Communication (MIKON-2004)*, Vol. 1, May 17–19, 2004, pp. 92–96.

[116] Pace, P. E., *Advanced Techniques for Digital Receivers*, Norwood, MA: Artech House, 2000.

[117] Mitola, J., "The Software Radio Architecture," *IEEE Communications Magazine*, Vol. 33, May 1995, pp. 26–38.

[118] Kulpa, K. S., Z. Czekala, and M. Smolarczyk, "Long-Integration-Time Noise Surveillance Radar," *First International Workshop on the Noise Radar Technology NRTW-2002*, September 18–20, 2002, pp. 238–243.

[119] Bará, J., et al., "Remote Sensing of Hydrometeors by Means of Interferometric Radiometry: Theory and Experimental Results," *Radio Science*, Vol. 35, No. 3, May-June 2000, pp. 799–812.

[120] Morabito, A. N., M. G. Meyer, and J. D. Sahr, "Improved Computational Performance for Distributed Passive Radar Processing Through Channelised Data," *IEE Proc. Radar, Sonar & Navig.*, Vol. 152, No. 3, June 3, 2005 pp. 179–184.

[121] Ellingson, S. W., "Capabilities and Limitations of Adaptive Canceling for Microwave Radiometry," *2002 IEEE International Geoscience and Remote Sensing Symposium (IGARSS '02)*, Vol. 3, June 24–28, 2002, pp. 1685–1687.

8

Examples of Noise Radars

In the previous chapters the theory of the noise radar was presented. Up to now there have not been any commercially available noise radars and the technology is not mature yet, but more examples of noise radar demonstrators have been shown. Some are still based on analog signal processing, becuase such an approach leads to cheap, simple solutions. However, the majority of present-day examples are based on modern digital technology.

Several experiments with noise radars have been carried out in the Radar Technology Laboratory at the Warsaw University of Technology. One such experiment, limited synthetic aperture radar, was described in Chapter 6. The structure of the experimental noise radar used in most of the experiments at the Warsaw University of Technology is presented in Figure 8.1. The noise source is an arbitrary waveform generator. The digital memory of the generator stores the real and imaginary parts of the baseband complex signal (IQ). Values of the sample are read from the memory and converted to the analog form using two digital analog converters (DACs) with a 14-bit resolution in this case. The IQ signals from the converters are then lowpass-filtered to remove the high-frequency component (higher than the Nyquist frequency) and upconverted by mixing them with the carrier frequency. The signal from the generator is split into two paths: the transmit signal and the reference signal. The dominant part of the generated power is transmitted through the transmit antenna. A fraction of the signal, taken from the low power output of a 20-dB splitter (directional coupler), is fed to the first receiver channel as a reference used in signal processing. The signal from the receive antenna is amplified in a low-noise amplifier and fed to the second channel of the receiver. In the receiver, both channels are downconverted using a common local oscillator (LO). The use of the same LO signal guarantees frequency and phase coherence of the receiver. The received signals after downconversion are lowpass-filtered and digitized using a common

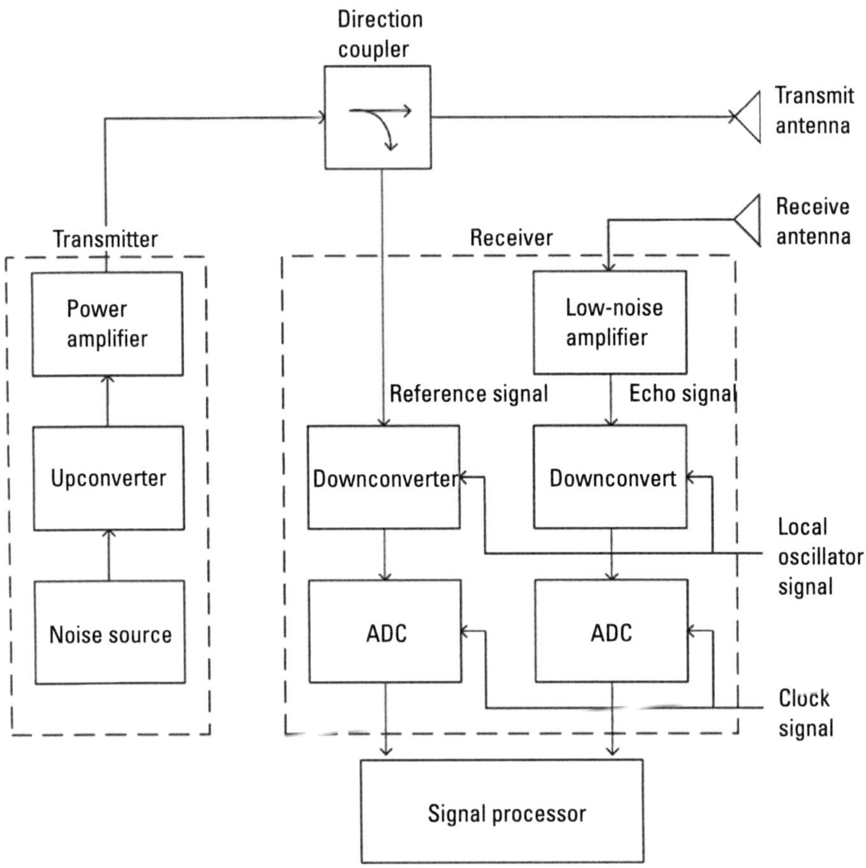

Figure 8.1 Block structure of a digital noise radar.

ADC clock. After digitizing, the signals are passed to the digital IQ, demodulated, and two IQ data streams (14 bits in this case) are passed to the signal processing unit. This unit can either store the data to the internal memory or hard drives matrices or process it in real time. For this purpose, different commercial-off-the-shelf (COTS) hardware was used. The examples of such hardware are presented in Figures 8.2 and 8.3.

In Figure 8.2 the single channel noise radar hardware is presented. The arbitrary waveform generator can store about 100 million complex samples in its internal memory and the vector signal analyzer can also store a similar number of samples with an instantaneous bandwidth of 36 MHz. The range resolution in the single frequency band is then at the level of 5m. A better resolution can be obtained using multiple bands. The length of the recording is equal to the sampling interval multiplied by the number of samples, so depending on the chosen bandwidth, it can vary from 2 seconds to 2 minutes. After the data

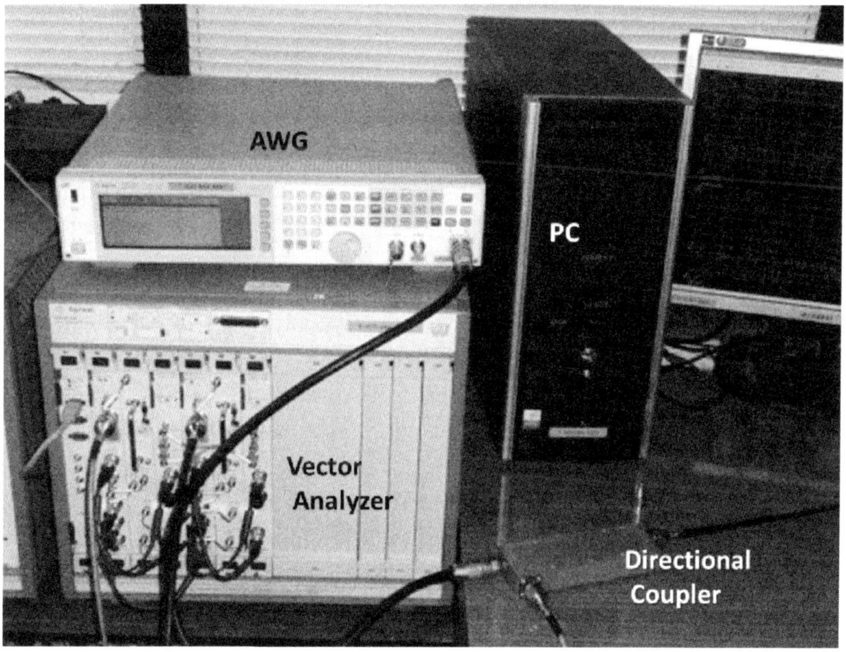

Figure 8.2 The Agilent arbitrary signal generator and two-channel vector signal analyzer used for the noise radar demonstrator.

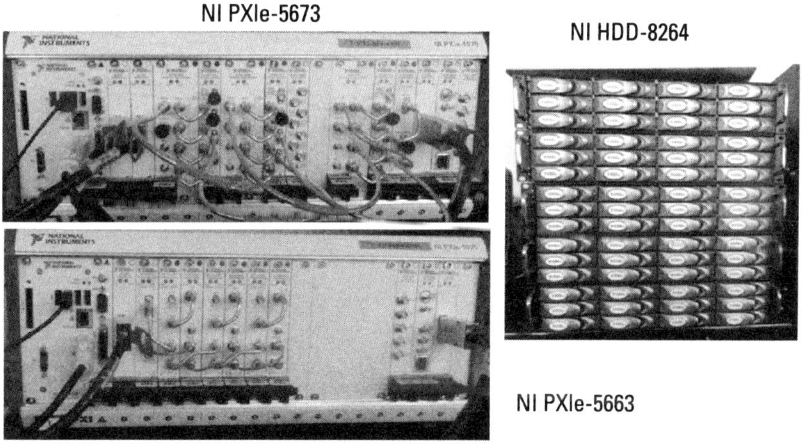

Figure 8.3 The National Instrument three-channel generator (upper-left) and three-channel signal analyzer (lower-left) with hard disk matrix (right) used for the noise radar and MIMO noise radar experiments.

acquisition period, it is necessary to upload the recorded data to the computer controlling the radar demonstrator, which takes up to 15 minutes.

The second hardware setup is presented in Figure 8.3. The hardware consists of a three-channel signal generator and a three-channel vector signal analyzer. The instantaneous bandwidth of the whole system is 50 MHz, which is equivalent to the monostatic range resolution of 3m. Both the generator and vector analyzer can be connected directly to the hard disk matrix so it is possible to stream recorded data directly to the hard drives and also get the recorded data to the generator. As a result, it is possible to record the three-channel data (and also to generate three channel data) continuously for more than 24 hours. The time of the recording (and playing) is now limited only by the capacity of the hard disk matrix.

The first experiments on noise radar concerned the detection of moving targets. The first targets of opportunity were cars on the street near the Warsaw University of Technology. The measurements were taken from a balcony of the Radar Technology Lab [1]. The measurement scenario together with the antenna setup is presented in Figure 8.4. As the transmit and receive antennas, two log-period wideband antennas for the 800 to 2,100-MHz band were used. The transmit antenna was placed on the right side and the receive antenna was

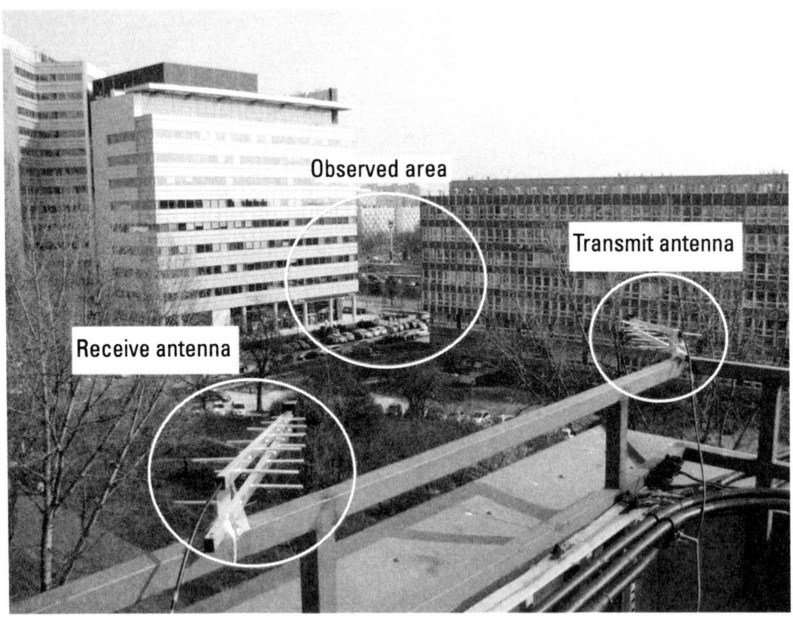

Figure 8.4 The first measurement using the noise radar demonstrator on the balcony of the Radar Technology Laboratory, Warsaw University of Technology. The targets of opportunity were cars in the parking lot and cars on the street.

placed on the left side. The area of interest was the street visible between the two buildings and the parking lot. The distance to the street was approximately 250m. During the experiment cars on the road and cars in the parking lot were detected after the application of the ground clutter removal procedure [2]. After this success, more experiments were carried out. For the second experiment a field environment was chosen [3]. The observed area was a country road, and this time the noise radar demonstrator was equipped with two reflector parabolic antennas used for transmitting and receiving with a gain of 24 dBi. The picture of the noise radar demonstrator and cooperative target equipped with a GPS logger is presented in Figure 8.5. The radar carrier frequency of 2.45 GHz was chosen to work on the public band [2.4 GHz is the Industrial, Scientific and Medical (ISM) band]. The spectrum of the transmitted signal was Gaussian to avoid range sidelobes. The block length of 50 ms was chosen for analysis. The receive data was tapered using a Hamming time window. The cross-ambiguity function of the transmit signal is depicted in Figure 8.6.

Also in this case the calculation of the cross-ambiguity function between the transmit (reference) and receive signals shows only direct antenna coupling and strong ground clutter as shown in Figure 8.7. The crosstalk signal was at the level of 60 dB (arbitrary scale) and the noise floor was at the level of 10 dB. The noise cancellation procedure, described in Chapter 4, completely removed the crosstalk signal and ground clutter. As a result, the noise floor was reduced to the level of −35 dB (a 45-dB improvement) and all targets were visible as

Figure 8.5 Field test of the noise radar demonstrator working in the 2.4-GHz band.

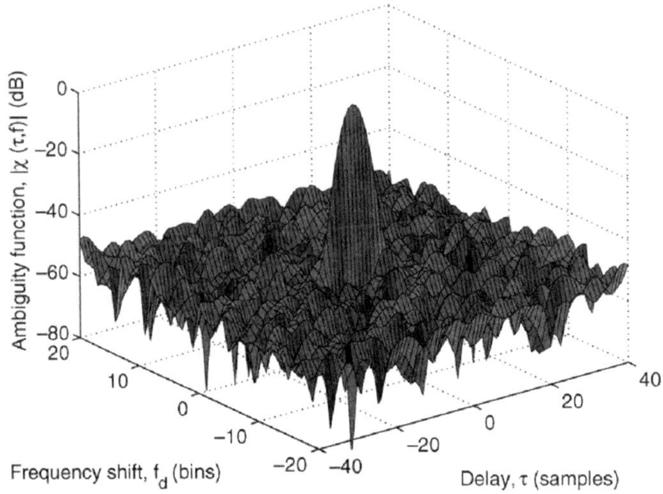

Figure 8.6 The transmitted signal cross-ambiguity function.

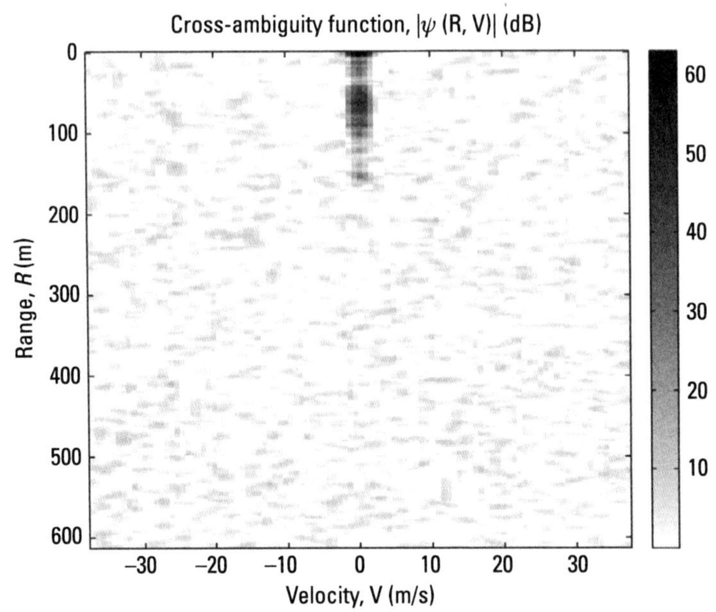

Figure 8.7 The result of signal processing in the car detection experiment; no clutter cancellation has been applied.

depicted in Figure 8.8. The target with the velocity of −10 m/s was the cooperating target equipped with a GPS logger. As can be seen in Figure 8.8, the cooperating target position and velocity measured by the noise radar is in very good agreement with the logged position and velocity. The targets at velocities

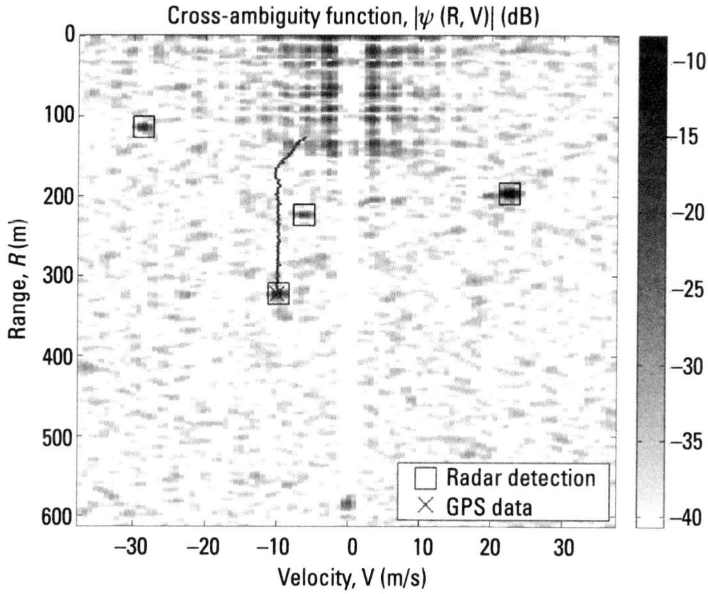

Figure 8.8 The result of signal processing in the noise radar after clutter cancellation.

-30 m/s and 25 m/s are noncooperative targets of opportunity and the target with the velocity of -5 m/s was a cyclist.

The maximum detection range was about 600m as only that part of the road was illuminated by the radar.

To verify if it is possible to have longer-range detection, the next experiment was focused on airplane detection. The Warsaw Airport is about 8 km away from the university campus and the flight path of the arriving airplanes is only 6 km from the Radar Technology Laboratory. Again noise illumination was used together with two reflector antennas. The carrier frequency was 2.45 MHz and output power was 20 dBm. The view from the balcony of the laboratory where both antennas were placed is presented in Figure 8.9. The picture was taken during signal recording and the landing airplane is clearly visible.

The result of the signal processing (including ground clutter cancellation) is presented in bistatic range-Doppler coordinates in Figure 8.10. The noise radar detected the landing aircraft at a range of 5.7 km. The target range velocity was only -27 m/s as the flight path is almost perpendicular to the line of sight. The detected position was compared with the ground truth obtained by the MODE-S receiver. Again both positions are in very good agreement.

The noise radar demonstrator was also used to measure ground reflectivity. In this case it was necessary to measure all polarimetric components: ground reflectivity in HH, VV, HV, and VH polarity. To perform the measurements, four antennas were used: two transmit (H and V polarized) and two receive

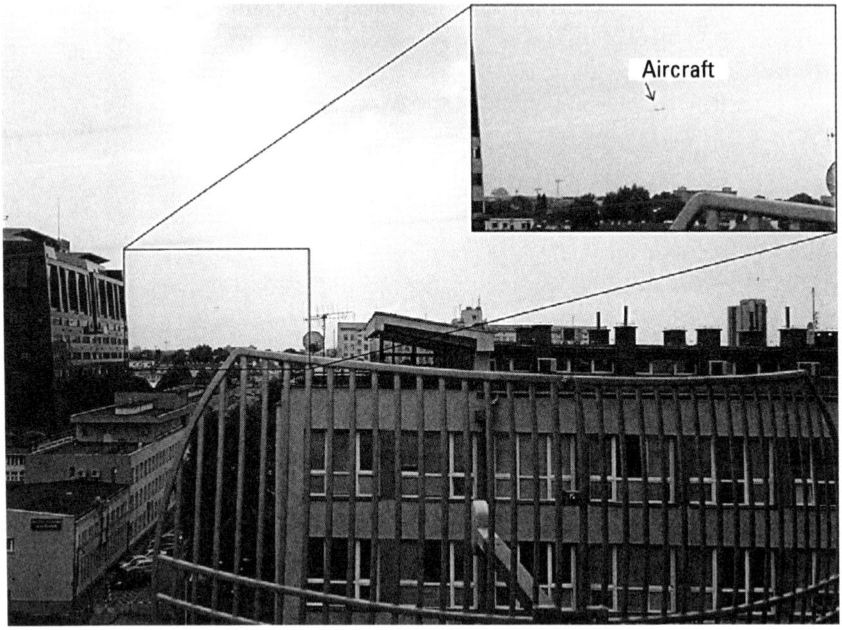

Figure 8.9 Aircraft detection experiment: view from the Radar Technology Laboratory balcony. The receive antenna of the noise radar demonstrator is visible at the bottom of the picture and the landing airplane is visible above the left corner of the antenna.

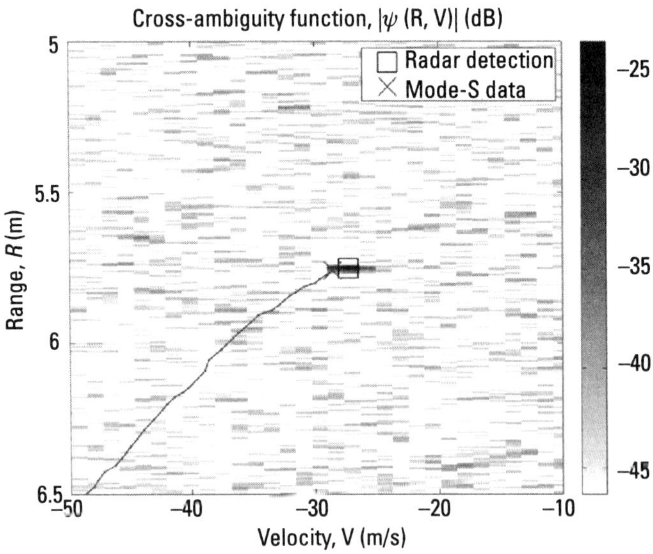

Figure 8.10 The result of signal processing. The radar noise detection marked by the square is very close to the parameters obtained from the MODE-S receiver. The solid line marks the target trajectory in the previous measurements.

(also H and V polarized). The two orthogonal noise signals were transmitted simultaneously and two received signal after amplification were recorded. For transmitting and recording a three-channel NI signal generator was used, and for receiving a three-channel signal analyzer was used. The block diagram of the scatterometer based on the noise radar demonstrator is shown in Figure 8.11. Two channels were used for signal transmission and recording, while a third channel was used for synchronization purposes. The picture of the experimental setup is presented in Figure 8.12. The four antennas were placed on top of a 12m deployable mast. All electronic devices were put into the camping trailer. To calibrate the scatterometer, the corner reflector of a known radar cross-section was used. The idea of scatterometric measurements is depicted in Figure 8.13. The radar signal of the selected bandwidth is emitted by transmit antennas. The reflected signal is collected by the receive antennas and the reflection originating from each range ring is calculated [4, 5]. The expected signal

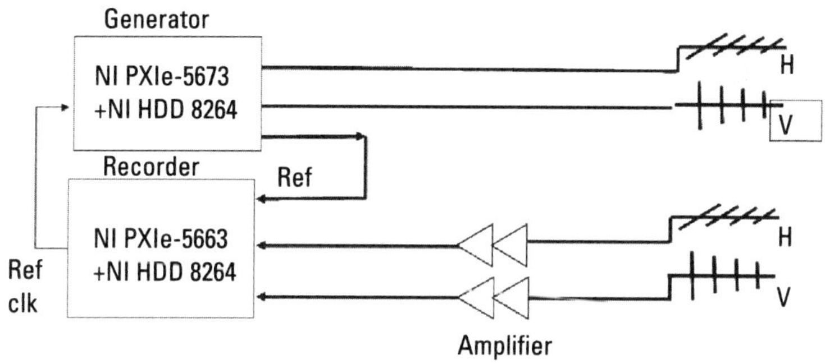

Figure 8.11 Block diagram of the polarimetric ground noise scatterometer.

Figure 8.12 Setup of the polarometric ground noise scatterometer (left), Tx/Rx antennas set (upper right), and a corner reflector used for calibration (bottom right).

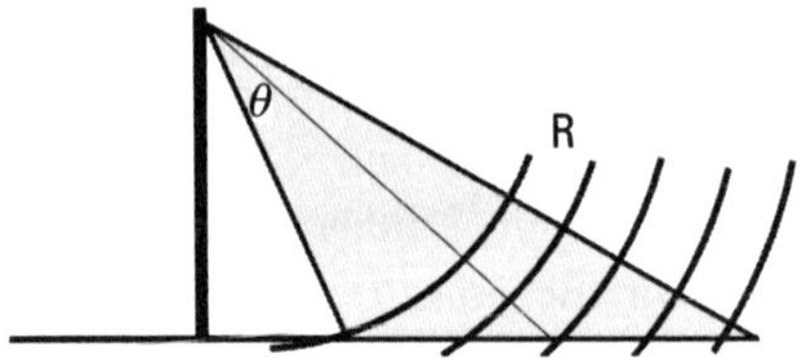

Figure 8.13 The idea of scatterometric measurements.

from the scatterometer is presented in Figure 8.14, while the measured values for each polarity are presented in Figure 8.15.

Intensive research on noise radar has also been carried out at the Military Technical University, Warsaw, Poland, in a laboratory led by Professor Bronislaw Stec. His group is working on both analog and digital-based noise radars. An example of the radar for a through-the-wall sensing application is presented in Figure 8.16. As it is for most continuous-wave radars, it is equipped with

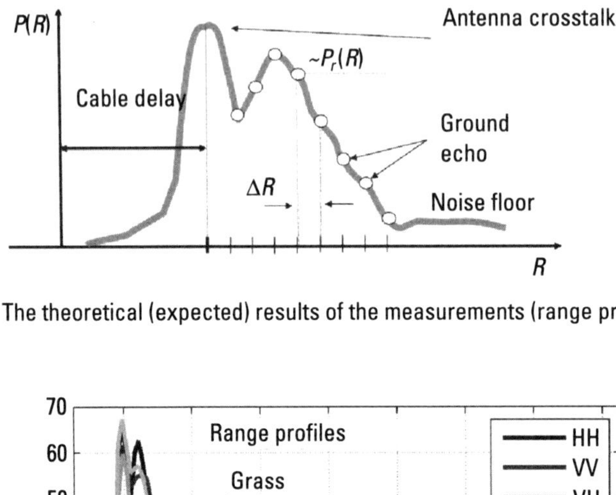

Figure 8.14 The theoretical (expected) results of the measurements (range profile).

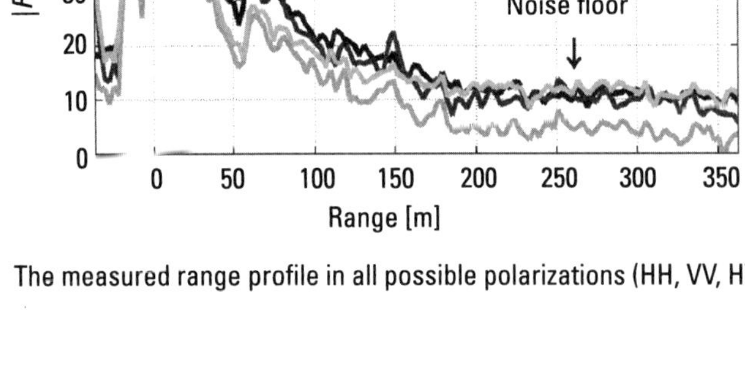

Figure 8.15 The measured range profile in all possible polarizations (HH, VV, HV, VH).

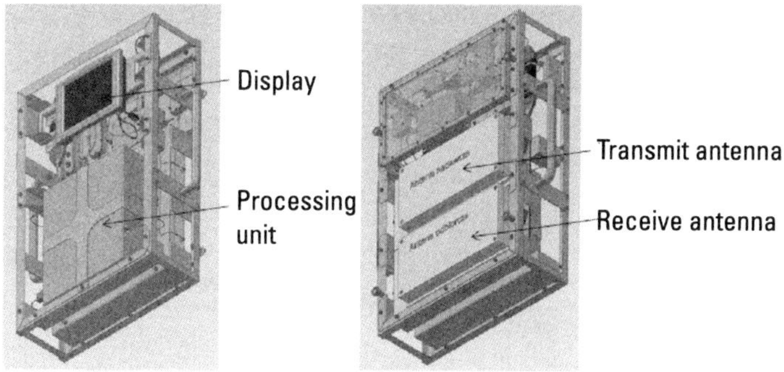

Figure 8.16 A through-the-wall noise radar.

separate transmit and receive antennas. The block diagram of the radar is presented in Figure 8.17.

The noise signal generated in the noise signal transmitter is radiated by the transmit antenna. Part of the transmit signal, taken from the coupler, is delayed by the tunable delay line and passed to two correlators producing in-phase (I) and quadrature (Q) signals. The signals after the multipliers (mixers) are lowpass-filtered using integrating analog circuits. An example of the output IQ signals is presented in Figure 8.18. Three different phases of the experiment are clearly visible. In period 1 the radar observed the scene when only a non-moving object was present. As it can be seen, the IQ signals are almost constant.

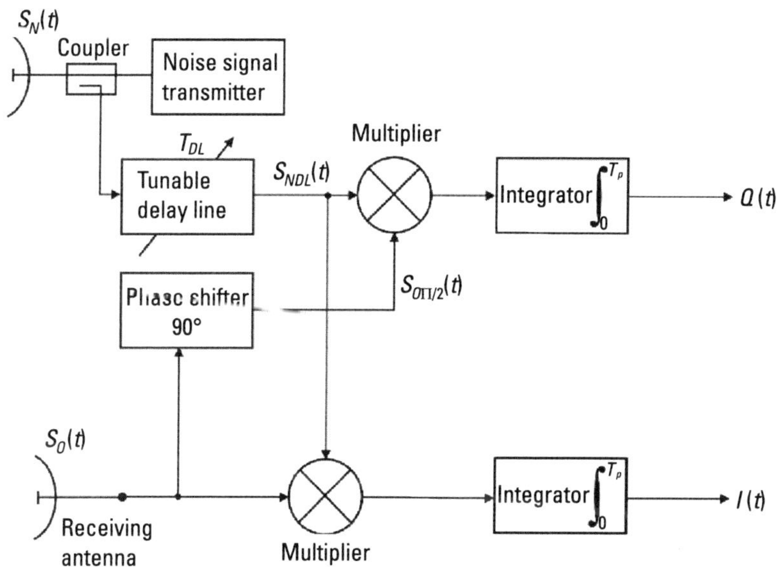

Figure 8.17 The block diagram of the noise radar with an analog correlator.

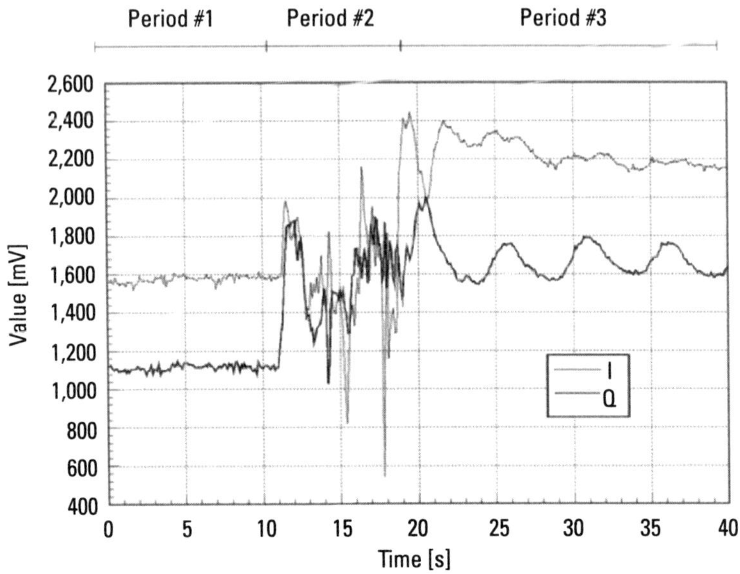

Figure 8.18 IQ output signals at the output of the analog correlator of the noise radar.

In period 2 a man entered the observation area, walked to a chair waving his hands, and sat on it. As can be seen, the IQ signals change rapidly in time. In period 3 the man was sitting still. The harmonic signal that originated from his chest movement due to his breathing can be observed. The distance between the radar and the man was about 6m. Similar signals were observed by Professor Stec's team when a man was hidden by a brick wall. Apart from Professor Stec, many scientists are working on the problem of through-the-wall object detection using wideband or ultrawideband noise radars [8–10].

Another very active scientist working in the noise radar technology area is Professor Konstantin Lukin from the Usikov Institute for Radiophysics and Electronics, at the National Academy of Sciences of Ukraine. His main area of interest is synthetic aperture noise radar. He has developed several noise radar demonstrators, described in his papers [12–18]. His synthetic aperture radar demonstrators are ground-based limited aperture radars. He has designed monostatic and multistatic systems working with different scanning concepts. One of the first was an arc-scanning SAR, where the SAR antenna was mounted at the end of a rotating arm, presented in Figure 8.19. In that particular case there was a single horn antenna used, while the radar worked in the pulse noise mode. The mechanism of the arm directed the antenna towards the scanning area during the whole 180° rotation of the arm. The example image obtained by the arc-SAR radar overlapping an aerial photo of Professor Lukin's laboratory is presented in Figure 8.20, where the building and surrounding trees are clearly visible.

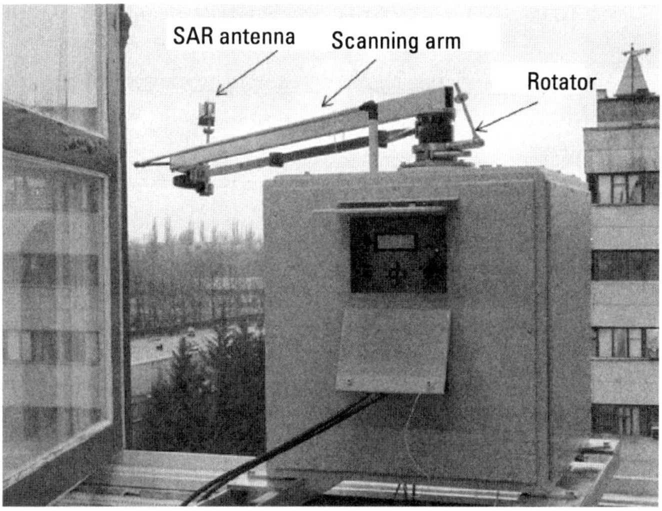

Figure 8.19 Photograph of the X-band noise arc-SAR based on a pulsed noise radar with coherent processing.

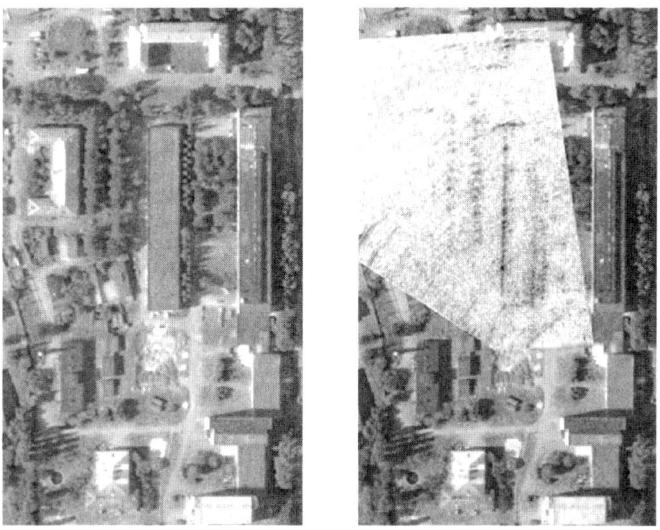

Figure 8.20 The SAR image generated using the X-band noise arc-SAR overlapped with Professor Lukin's laboratory aerial image (right) and the full aerial image of the laboratory (left).

Another interesting example of his noise SAR is presented in Figure 8.21. It is an illustration of an interferometry campaign held in Kiev, Ukraine, to measure the displacement of the bell tower during the day and night [11]. In this case a Ka-band, continuous-wave noise limited aperture SAR radar was used. The antenna was not in motion, but inside the waveguide antenna a moving

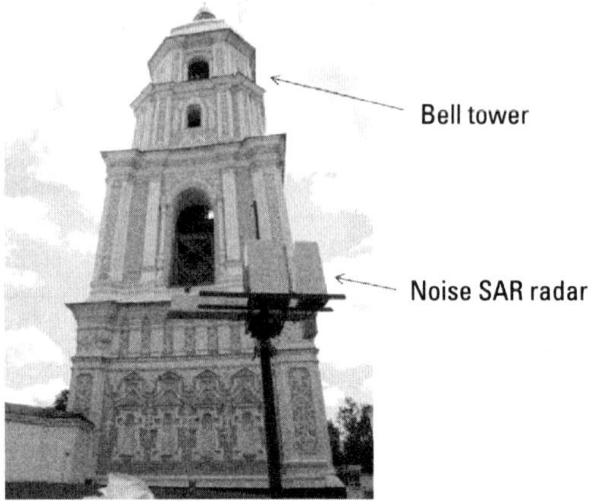

Figure 8.21 Photograph of the St. Sophia bell tower and the Ka-band noise SAR.

slot was placed. The movement of the slot makes a horizontal shift of the radiation center of the antenna, and this is why the SAR image could be obtained.

The SAR image of the bell tower is presented in Figure 8.22. The shape of the tower can be clearly seen in this picture. During the measurement campaign,

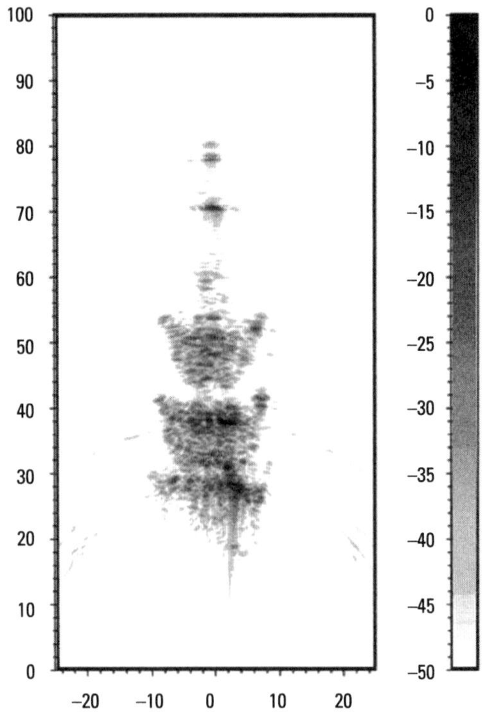

Figure 8.22 The SAR image of the bell tower generated using the Ka-band noise SAR.

several SAR images were taken, and these images were further used for the creation of interferograms (plots of phase changes between complex images). Two examples of interferograms are presented in Figure 8.23. The left interferogram was taken during nighttime, with almost constant ambient temperature. The shifts of the tower are very small, corresponding to changes of phase within 10°. The right image presents the interferogram taken during sunrise. The phase difference is now much higher, at the level of 40°. Further analysis shows that the phase measurement accuracy was at the level of 3°. In the Ka-band it corresponds to a measurement accuracy of the radial shifts at the level of 0.03 mm.

The example presented shows that noise radar technology is the subject of intensive research. Many scientists in different countries like the United States, Canada, the United Kingdom, Germany, Ukraine, Poland, Iran, South Korea, and many more are conducting research in this area, and in the near future noise radars will probably become commercially available.

All the technology presented is based on microwave devices, fast DAC and ADC, and an expensive signal processing unit. At first glance, it seems that only rich laboratories can make research in this area, but it has been shown that it is possible to carry out research in this area using acoustic signals instead of

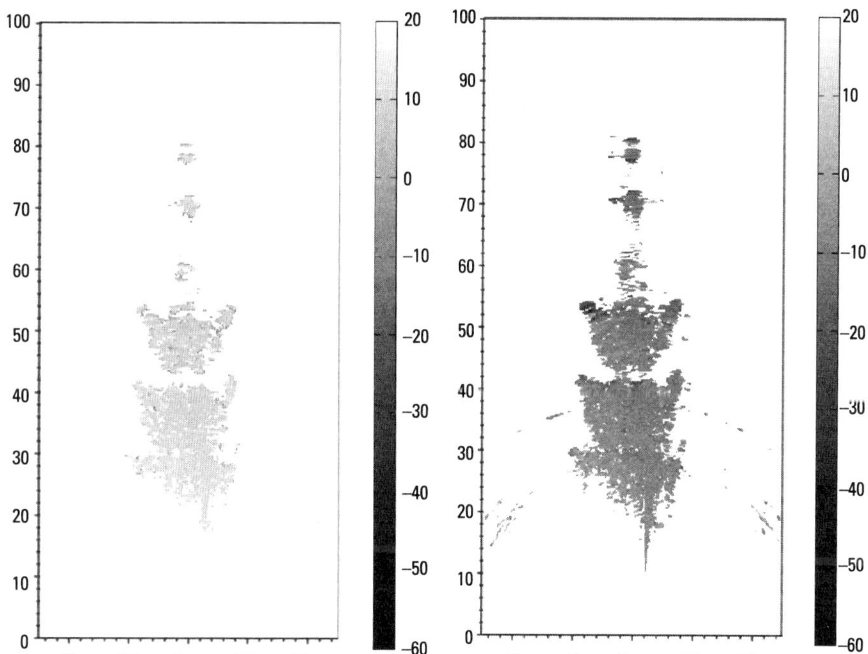

Figure 8.23 Differential SAR interferograms of the bell tower generated using the Ka-band noise SAR. Temperature variations in the scene during the night (left) and during sunrise (right).

microwaves [20, 21]. As it was also shown, it is possible to construct an interesting sensor and investigate noise radar-related problems using a classical personal computer equipped with a sound card, loudspeakers, and a microphone. To obtain a long detection range, acoustic parabolic mirrors can be used, and satellite TV dishes are also very useful for this purpose.

References

[1] Malanowski, M., et al., "Preliminary Results of Noise Radar Experiments," *Proceedings of SPIE: Photonics Applications in Astronomy, Communications, Industry, and High-Energy Physics Experiments,* 2009, pp. 7502OW-1–75020W-5.

[2] Kulpa, K. S., and Z. Czekała, "Ground Clutter Suppression in Noise Radar," *Proc. Int. Conf. RADAR 2004,* Toulouse, France, October 18–22, 2004, p. 236

[3] Malanowski, M., and K. Kulpa, "Detection of Moving Targets with Continuous-Wave Noise Radar: Theory and Measurements," *IEEE Transactions on Geoscience and Remote Sensing,* Vol. 50, No. 9, September 2012, pp. 3502–3509.

[4] Maslikowski, L., et al., "A Radar Demonstrator for Ground-Based L-Band Ground Reflectivity Measurements," *2012 9th European Radar Conference (EuRAD),* October 31–November 2, 2012, pp. 119–122.

[5] Łukasz, M., et al., "Preliminary Results of Ground Reflectivity Measurements Using Noise Radar," *Proceedings of SPIE: Photonics Applications in Astronomy, Communications, Industry, and High-Energy Physics Experiments, SPIE,* Vol. 8008, 2011, pp. 80081V-1-80081V-7.

[6] Susek, W., and B. Stec, "Broadband Microwave Correlator of Noise Signals," *Metrology and Measurement Systems, Polish Academy of Sciences,* Vol. XVII, No. 2, 2010, pp. 289–298.

[7] Susek, W., and B. Stec, "Noise Radar with Broadband Microwave Ring Correlator," *Proceedings of SPIE: 8021,* Radar Sensor Technology XV, 80211L, June 2011.

[8] Lai, C. -P., R. Qing, and R. M. Narayanan, "Hilbert-Huang Transform (HHT) Analysis of Human Activities Using Through-Wall Noise Radar," *International Symposium on Signals, Systems and Electronics 2007 (ISSSE '07),* Ecole Polytechnique de Montreal, Montreal, QC, Canada, July 30–August 2, 2007, pp. 115–118.

[9] Wu, J., et al., "Compressive Sensing for Sense-Through-Wall UWB Noise Radar Signal," *2011 6th International ICST Conference on Communications and Networking in China (CHINACOM),* August 17–19, 2011, pp. 979–983.

[10] Narayanan, R. M., "Through Wall Radar Imaging Using UWB Noise Waveforms," *IEEE International Conference on Acoustics, Speech and Signal Processing, 2008 (ICASSP 2008),* March 31–April 4, 2008, pp. 5185–5188.

[11] Lukin, K., et al., "Monitoring of St. Sophia Cathedral Interior Using Ka-Band Ground Based Noise Waveform SAR," *European Radar Conference 2009 (EuRAD 2009),* September 30–October 2, 2009, pp. 215–217.

[12] Lukin, K. A., et al., "Noise Waveform SAR for 2D and 3D Imaging," *2010 11th International Radar Symposium (IRS),* June 16–18, 2010, pp. 1–4.

[13] Lukin, K. A., et al., "Ka-Band Ground Based Noise Waveform SAR," *The Sixth International Kharkov Symposium on Physics and Engineering of Microwaves, Millimeter and Submillimeter Waves and Workshop on Terahertz Technologies 2007 (MSMW '07)*, Vol. 1, June 25–30, 2007, pp. 159–164.

[14] Lukin, K. A., et al., "Ka-Band Bistatic Ground-Based Noise Waveform SAR for Short-Range Applications," *IET Radar, Sonar & Navigation*, Vol. 2, No. 4, August 2008, pp. 233–243.

[15] Lukin, K., et al., "2D and 3D Imaging Using S-Band Noise Waveform SAR," *2011 3rd International Asia-Pacific Conference on Synthetic Aperture Radar (APSAR)*, September 26–30, 2011, pp. 1–4.

[16] Lukin, K., et al., "L-Band Stepped Frequency Noise SAR on the Basis of Arbitrary Waveform Generator," *2011 3rd International Asia-Pacific Conference on Synthetic Aperture Radar (APSAR)*, September 26–30, 2011, pp. 1–4.

[17] Lukin, K., A. Mogyla, and P. Vyplavin, "Ground Based Noise Waveform SAR with Static Antenna Array and Single Channel Receiver," *2008 7th European Conference on Synthetic Aperture Radar (EUSAR)*, June 2–5, 2008, pp. 1–3.

[18] Lukin, K. A., A. A. Mogila, and P. L. Vyplavin, "Ground-Based Noise Waveform SAR and Differential Interferometry for Remote Monitoring of Large Objects," *The Sixth International Kharkov Symposium on Physics and Engineering of Microwaves, Millimeter and Submillimeter Waves and Workshop on Terahertz Technologies 2007 (MSMW '07)*, Vol. 1, June 25–30, 2007, pp. 445–447.

[19] Lukin, K., and A. Mogyla, "Static Nonswitchable Antenna Array for Ground-Based Noise Waveform SAR: A Novel Concept for SAR Imaging," *International Radar Symposium 2006 (IRS 2006)*, May 24–26, 2006, pp. 1–4.

[20] Sinitsyn, R. B., and F. J. Yanovsky, "UWB Noise Acoustic Atmospheric Radar," *2010 5th International Conference on Ultrawideband and Ultrashort Impulse Signals (UWBUSIS)*, September 6–10, 2010, p. 48.

[21] Axelsson, S. R. J., "Acoustic Random Noise Radar Using Ultra Wide Band Waveforms," *International Radar Symposium 2006 (IRS 2006)*, May 24–26, 2006, pp. 1–4.

List of Symbols

A	amplitude of the signal
A_d	amplitude of the direct (crosstalk) signal
B	bandwidth
c	velocity of light, 299,792,458 [ms^{-1}] (in vacuum)
C	crosstalk coefficient
D_o	detectability factor
E_T	transmitting pulse energy
F	carrier frequency
f_d	Doppler frequency
G	gravity constant 6.672×10^{-11} [$m^3kg^{-1}\ s^{-2}$]
G_R	receiving antenna gain
G_T	transmitting antenna gain
h	Planck's constant 6.626068×10^{-34} [m2kg/s]
H	height
k	Boltzmann's constant $1.3806505 \times 10^{-23}$ [$J\ K^{-1}$]
L	total insertion losses
L_a	antenna aperture length

P_d	direct signal power
P_N	thermal noise power at the input of the receiver
P_R	received (echo) signal power
PRF	pulse repetition frequency
P_T	transmitted power (peak)
P_{Tm}	mean transmitted power
R	distance from radar to a target
R_E	the Earth's radius
$R_{\mathbf{max}}$	maximum value of detection range
$R_{\mathbf{min}}$	minimum value of detection range
S_o	effective target cross-section
S_R	effective receiving antenna surface
S_T	effective transmitting antenna surface
t	time
t_i	signal integration time
T_{ob}	target observation time
t_p	pulse width
T_R	the effective receiver noise temperature
t_s	scan time
T_s	sampling time (interval)
W_a	antenna aperture width (height)
x_R	received signal in the time domain
X_R	received signal in the frequency domain
x_T	transmitted signal in the time domain
X_T	transmitted signal in the frequency domain
v	velocity

ξ_t	thermal noise signal
λ	wavelength
τ	time delay

Index

Recent Titles in the Artech House Radar Series

David K. Barton, Series Editor

Adaptive Antennas and Phased Arrays for Radar and Communications, Alan J. Fenn

Advanced Techniques for Digital Receivers, Phillip E. Pace

Advances in Direction-of-Arrival Estimation, Sathish Chandran, editor

Airborne Pulsed Doppler Radar, Second Edition, Guy V. Morris and Linda Harkness, editors

Bayesian Multiple Target Tracking, Lawrence D. Stone, Carl A. Barlow, and Thomas L. Corwin

Beyond the Kalman Filter: Particle Filters for Tracking Applications, Branko Ristic, Sanjeev Arulampalam, and Neil Gordon

Cognitive Radar: The Knowledge-Aided Fully Adaptive Approach, Joseph R. Guerci

Computer Simulation of Aerial Target Radar Scattering, Recognition, Detection, and Tracking, Yakov D. Shirman, editor

Design and Analysis of Modern Tracking Systems, Samuel Blackman and Robert Popoli

Detecting and Classifying Low Probability of Intercept Radar, Second Edition, Phillip E. Pace

Digital Techniques for Wideband Receivers, Second Edition, James Tsui

Electronic Intelligence: The Analysis of Radar Signals, Second Edition, Richard G. Wiley

Electronic Warfare in the Information Age, D. Curtis Schleher

Electronic Warfare Target Location Methods, Second Edition, Richard A. Poisel

ELINT: The Interception and Analysis of Radar Signals, Richard G. Wiley

EW 101: A First Course in Electronic Warfare, David Adamy

EW 102: A Second Course in Electronic Warfare, David Adamy

EW 103: Tactical Battlefield Communications Electronic Warfare, David Adamy

Fourier Transforms in Radar and Signal Processing,Second Edition, David Brandwood

Fundamentals of Electronic Warfare, Sergei A. Vakin, Lev N. Shustov, and Robert H. Dunwell

Fundamentals of Short-Range FM Radar, Igor V. Komarov and Sergey M. Smolskiy

Handbook of Computer Simulation in Radio Engineering, Communications, and Radar, Sergey A. Leonov and Alexander I. Leonov

High-Resolution Radar, Second Edition, Donald R. Wehner

Introduction to Electronic Defense Systems, Second Edition, Filippo Neri

Introduction to Electronic Warfare, D. Curtis Schleher

Introduction to Electronic Warfare Modeling and Simulation, David L. Adamy

Introduction to RF Equipment and System Design, Pekka Eskelinen

Introduction to Modern EW Systems, Andrea De Martino

The Micro-Doppler Effect in Radar, Victor C. Chen

Microwave Radar: Imaging and Advanced Concepts, Roger J. Sullivan

Millimeter-Wave Radar Targets and Clutter, Gennadiy P. Kulemin

Modern Radar Systems, Second Edition, Hamish Meikle

Modern Radar System Analysis, David K. Barton

Modern Radar System Analysis Software and User's Manual, Version 3.0, David K. Barton

Monopulse Principles and Techniques, Second Edition, Samuel M. Sherman and David K. Barton

MTI and Pulsed Doppler Radar with MATLAB®, Second Edition, D. Curtis Schleher

Multitarget-Multisensor Tracking: Applications and Advances Volume III, Yaakov Bar-Shalom and William Dale Blair, editors

Principles of High-Resolution Radar, August W. Rihaczek

Principles of Radar and Sonar Signal Processing, François Le Chevalier

Radar Cross Section, Second Edition, Eugene F. Knott, et al.

Radar Equations for Modern Radar, David K. Barton

Radar Evaluation Handbook, David K. Barton, et al.

Radar Meteorology, Henri Sauvageot

Radar Reflectivity of Land and Sea, Third Edition, Maurice W. Long

Radar Resolution and Complex-Image Analysis, August W. Rihaczek and Stephen J. Hershkowitz

Radar Signal Processing and Adaptive Systems, Ramon Nitzberg

Radar System Analysis, Design, and Simulation, Eyung W. Kang

Radar System Analysis and Modeling, David K. Barton

Radar System Performance Modeling, Second Edition, G. Richard Curry

Radar Technology Encyclopedia, David K. Barton and Sergey A. Leonov, editors

Radio Wave Propagation Fundamentals, Artem Saakian

Range-Doppler Radar Imaging and Motion Compensation, Jae Sok Son, et al.

Robotic Navigation and Mapping with Radar, Martin Adams, John Mullane, Ebi Jose, and Ba-Ngu Vo

Signal Detection and Estimation, Second Edition, Mourad Barkat

Signal Processing in Noise Waveform Radar, Krzysztof Kulpa

Space-Time Adaptive Processing for Radar, J. R. Guerci

Special Design Topics in Digital Wideband Receivers, James Tsui

Theory and Practice of Radar Target Identification, August W. Rihaczek and Stephen J. Hershkowitz

Time-Frequency Signal Analysis with Applications, Ljubiša Stanković, Miloš Daković, and Thayananthan Thayaparan

Time-Frequency Transforms for Radar Imaging and Signal Analysis, Victor C. Chen and Hao Ling

For further information on these and other Artech House titles, including previously considered out-of-print books now available through our In-Print-Forever® (IPF®) program, contact:

Artech House
685 Canton Street
Norwood, MA 02062
Phone: 781-769-9750
Fax: 781-769-6334
e-mail: artech@artechhouse.com

Artech House
16 Sussex Street
London SW1V HRW UK
Phone: +44 (0)20 7596-8750
Fax: +44 (0)20 7630-0166
e-mail: artech-uk@artechhouse.com

Find us on the World Wide Web at: www.artechhouse.com